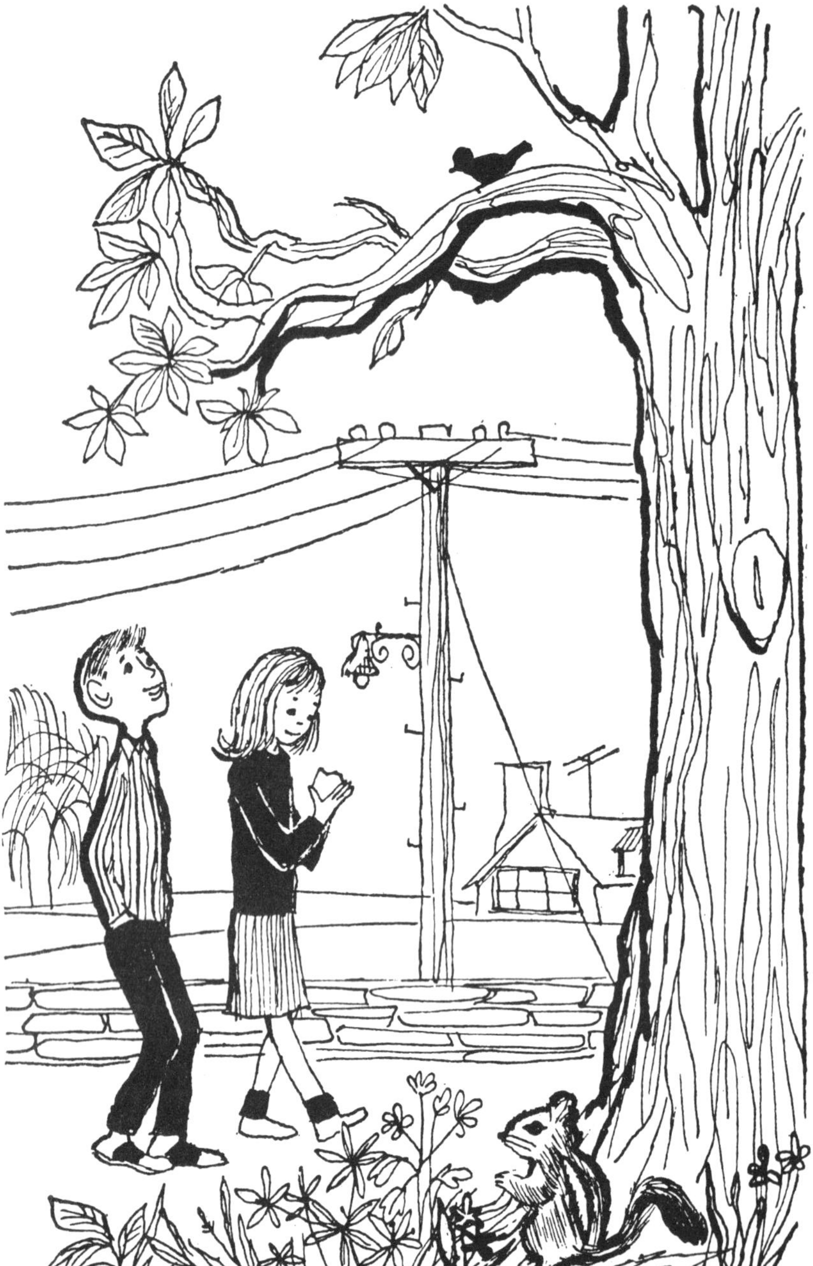

OUTDOOR SCIENCE PROJECTS FOR YOUNG PEOPLE

by George Barr

Illustrated by Jeanne Bendick

DOVER PUBLICATIONS, INC.
New York

Copyright © 1959 by George Barr and Jeanne Bendick.
All rights reserved under Pan American and International Copyright Conventions.

Published in Canada by General Publishing Company, Ltd., 30 Lesmill Road, Don Mills, Toronto, Ontario.
Published in the United Kingdom by Constable and Company, Ltd., 3 The Lanchesters, 162–164 Fulham Palace Road, London W6 9ER.

This Dover edition, first published in 1991, is an unabridged republication of the work first published by the McGraw-Hill Book Company, New York, 1959, under the title *Young Scientist Takes a Walk*.

Manufactured in the United States of America
Dover Publications, Inc., 31 East 2nd Street, Mineola, N.Y. 11501

Library of Congress Cataloging-in-Publication Data

Barr, George, 1907–
  [Young scientist takes a walk]
  Outdoor science projects for young people / by George Barr ; illustrated by Jeanne Bendick.
    p.   cm.
  Reprint. Originally published: Young scientist takes a walk. New York : McGraw-Hill, c1959.
  Includes bibliographical references and index.
  Summary: Noises, odors, sounds, and colors of city and country provide the basis of scientific inquiry and exploration as readers learn to recognize birds, animal prints, constellations, plants, metals, and stones.
  ISBN 0-486-26855-1 (pbk.)
  1. Scientific recreations—Juvenile literature. 2. Natural history—Juvenile literature. 3. Science—Philosophy—Juvenile literature. [1. Scientific recreations. 2. Natural history.] I. Bendick, Jeanne, ill. II. Title.
Q164.B344   1991
500—dc20   91-4342
                                                               CIP
                                                               AC

CONTENTS

Introduction	7
About Sidewalks and Streets	9
Lending an Ear to Sounds	22
Becoming Aware of Odors	31
Recognizing Building Stones	40
Metals Seen in the Neighborhood	51
Nothing Lasts Forever	62
How to Look at Trees	73
Plants along the Way	84
Animal Neighbors	97
Insects Galore!	107
Getting Acquainted with Birds	119
The Sky in the Daytime	133
Observing the Night Sky	143
Other Books for the Observant Walker	156
Index	158

Note: At the end of every chapter will be found More to Find Out

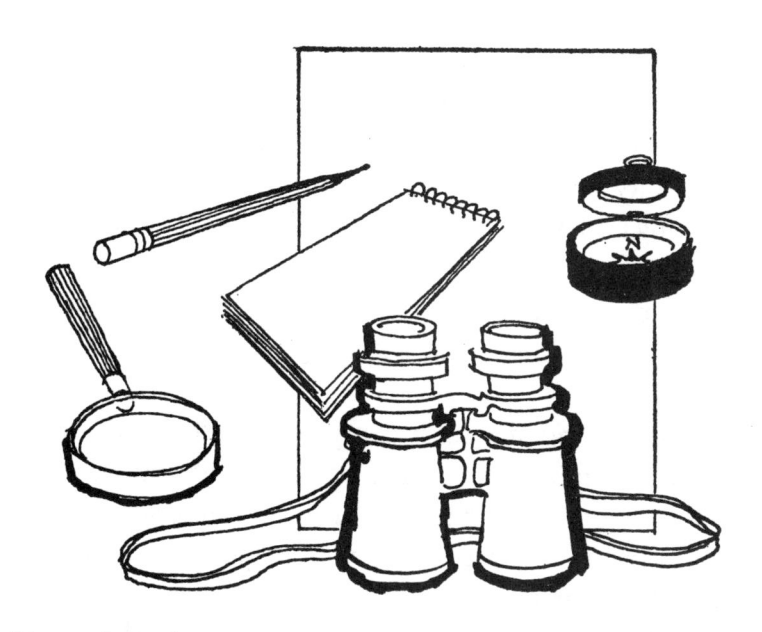

Introduction

A walk through your neighborhood can be an exciting adventure. It is an opportunity to exercise your powers of observation and use your scientific knowledge to explain the many things around you which illustrate some fact or principle in science.

It makes little difference whether the walking takes place in bustling cities or small towns, along busy avenues or dirt roads, past lawns or through meadows and woods. There is always something of interest to capture your attention.

Along the way, noises and odors fill the air. Insects, birds, and other animals make their presence known in many ways. Trees, shrubs, and colorful flowers are always fascinating objects to stop and examine. It is fun to recognize stones and metals and to see how they are used. On all sides you can also observe how the forces of Nature are at work building and destroying.

The day sky, as well as the night sky, is full of wonderful things to view on a walk. It is refreshing to see changeable clouds, rainbows, and colorful sunsets. It is a pleasure to observe the moon, constellations, and planets. You may even catch a glimpse of a shooting star!

Come equipped with a compass, a pad, pencil, and a small magnifying glass. As a special treat you may wish to take binoculars for seeing birds. A camera is another useful device for recording your experiences to share later with your friends.

This book directs your attention to many things you will encounter on your walks near your home.

Now—let's go exploring!

ABOUT SIDEWALKS AND STREETS

Much of your walking is done on sidewalks and streets. Because they are so commonplace we take them for granted, yet they can be the source of many interesting scientific observations.

Most of us hardly ever give a thought to the fact that civilization has come a long way from the simple, trampled-grass cowpath to the complicated, paved street of today.

Has it ever occurred to you how many useful and important functions a city street performs?

Among other things, it provides a hard, durable surface for the movement of heavy vehicles and many people. Surveyors and civil engineers have slanted it so that it is not too steep for traffic to move easily.

Under the streets and sidewalks is a maze of pipes for water, for cooking gas, and for sewage disposal. There are also electric cables and telephone wires.

CURB
GUTTER
CROWN

Look at the center of the street. You probably will find that the crown, as it is called, is visibly higher than the section near the gutter which is alongside the curb.

A street or sidewalk may frequently seem perfectly level and flat, but there is always enough of a slant to allow the water to run down to a sewer.

Sometimes after a very heavy downpour you may see a place where the water is not removed fast enough. This causes a flood area which may damage property in nearby basements.

When walking on a dirt road, see how the Highway Department has slanted it so that a heavy rain does not wash out the road. Observe the drainage ditches on the sides. Notice especially how the low spots or the dips in the road have been treated to prevent lakes from forming there during rainstorms.

CONCRETE　　　　　ASPHALT

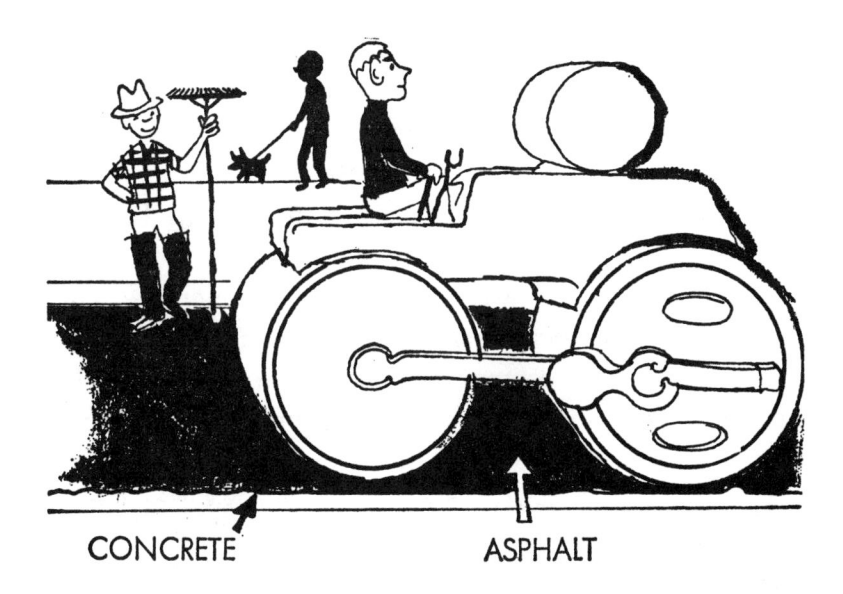

Did you ever stop to watch a city street being paved? First, a heavy concrete layer is poured. When this foundation hardens in about a week, black asphalt is poured over it and pressed down smoothly by a heavy roller.

Asphalt used to be obtained from natural asphalt pits and lakes in Trinidad, British West Indies. Today almost all of this wonderful road-building material comes from the oil and gasoline refineries in the United States. It is the tarlike substance left after most other chemicals have been boiled out of petroleum. To make it thicker and harder for pavements, rock dust and other ingredients are added.

Asphalt is cheap, easy to work with, and can withstand heavy traffic, as well as all kinds of weather. It is able to expand and contract in the summertime and in the wintertime without too much cracking. No wonder it is used so widely.

Sometimes you may see roads which have small chunks of stone bonded together by asphalt and smoothed down by heavy rollers. These are called macadam (muh-KAD-um) roads, named after a famous road builder, John MacAdam, over 100 years ago.

A better, but more expensive surface is concrete. This hard, man-made rock is a mixture of cement, sand, gravel, and water. The cement is manufactured in mills by heating certain clays and shale with limestone in special rotating furnaces. When ground up, powder-fine, this material is called Portland cement. It is sold in heavy paper bags weighing 94 pounds and has a volume of 1 cubic foot. You can see them piled up at construction jobs.

When this cement is mixed with water until it forms a thick paste, it will harden into a very hard mass. It does this by forming tiny crystals which intertwine about each other. If sand and gravel are added during the mixing operation, the concrete thus formed will have tremendous mechanical strength.

Try to obtain very small quantities of these three ingredients from someone who is working with concrete.

MIX
1 PART CEMENT
2 PARTS SMALL PEBBLES
2 PARTS SAND
ADD WATER AND MIX
POUR INTO A MOLD

Make a mixture of 1 part cement, 2 parts sand (not beach sand, which may contain salt), and 2 parts small pebbles. Add water and mix everything to form a thick paste. Pour it into a mold made from an empty milk carton. Allow about a week for it to harden. During this time it should be kept covered with a moist cloth or newspaper to prevent cracking.

Since concrete expands on hot summer days, concrete roads frequently have expansion spaces between sections which are filled with tar.

Some dirt roads in the country are made with a surface of gravel or clay soil. Sometimes a special heavy oil or light asphalt is used to hold the soil and stones together on the surface.

OIL-SPREADING TRUCK

Have you ever wondered why cement sidewalks are marked off with evenly spaced lines? These are not only for beauty; they have a special purpose. If you look carefully you will see that these lines prevent cracks from spreading. Those large spaces you see filled with tar, allow the sidewalks to expand in warm weather without buckling.

Some cement sidewalks have small, evenly spaced depressions over the entire surface. These marks were left by the heavy, dimpled paper which was laid over the sidewalks while the cement was hardening. It prevented too rapid drying, which might have produced cracks.

Sidewalks are damaged by expansion and contraction due to heat and cold, by heavy blows, by upheaval of soil caused by roots of trees and shrubs, by erosion, and other reasons.

THESE LINES PREVENT CRACKS FROM SPREADING

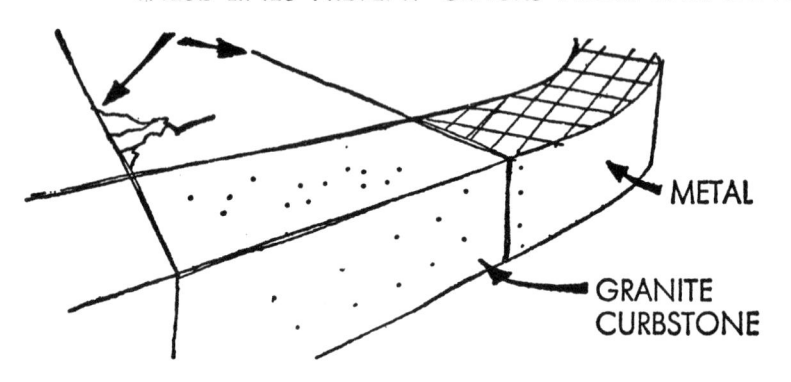

METAL

GRANITE CURBSTONE

Curbstones, which take much punishment from the wheels of automobiles, are usually made from some hard stone, such as granite. Corner curbstones need more protection, so the edges may have steel around them. Some corners are protected by solid iron plates.

The next time you walk on a curved road, notice that the road is banked; that is, it is higher on the outside of the curve. There is a good scientific reason for this protective design.

Try to recall when you were in a car making a sharp left turn and everybody moved toward the right. This is due to inertia, a tendency of objects moving in a certain direction to continue in that direction. When the moving car turned to the left, everything had a tendency to move in the original direction—straight ahead, with the result that you moved to the right side of the car in a most undignified manner.

This may even tip over a very fast-moving automobile. Raising the right side of the road supports the right wheels and prevents the car from tipping dangerously. Watch for these banked roads. Notice also that the faster the traffic is allowed to move, the sharper the road is banked.

Have you ever walked on a hot day and thought you saw a puddle of water shimmering far ahead in the middle of the road? When you came up close you found nothing there but the pavement. This "mirage" is caused by a layer of very warm air close to the hot road. Light from the sky is bent by this layer and sent to your eyes before it can be reflected from the road. This looks so much like rippling water that the eye is fooled.

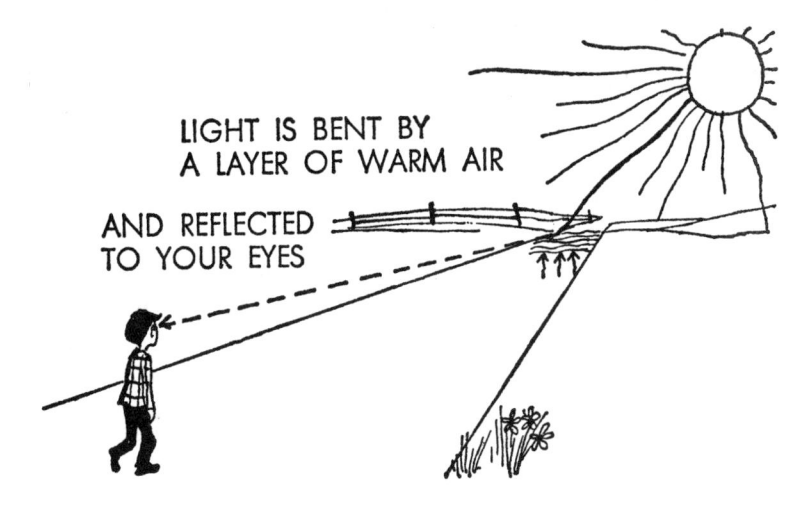

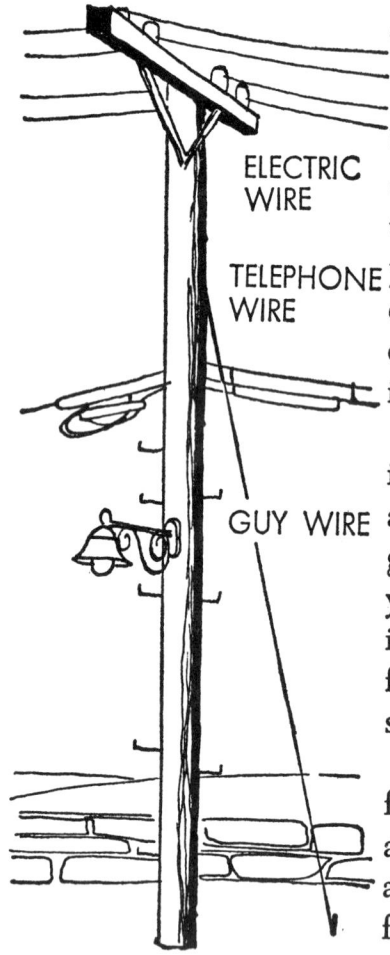

About electric poles

ELECTRIC WIRE

TELEPHONE WIRE

GUY WIRE

One of the most familiar objects on most sidewalks is the electric light pole, yet you can stand in front of one for a long time and pick out items of great scientific interest.

While it is true that today in many large cities the wires are being placed underground, it will be many years before the old-fashioned "telegraph pole" will fade from the American scene.

The reason is not hard to find. The tall straight poles are convenient workhorses and perform too many different tasks to be easily replaced.

Let us begin by finding out what kind of wood is used. Can a hard wood like maple or oak be used? Obviously not, because the lineman could not dig his climbing irons into hard wood.

Look at the holes in the pole made by the spikes of the climbing irons. Many modern poles are made of cedar because it is a soft wood and withstands the elements well, but does not rot in the ground as easily as other kinds of wood. Pine is also used, but only if it is soaked throughout by creosote (KREE-uh-sote) to prevent termites, rot, and water from destroying the part which is in the ground.

Can you see the black creosote oozing out of the wood, especially near the bottom of the pole? Sometimes it becomes a nuisance when it runs onto the sidewalk.

Does the pole have foothold hooks (also called pole steps) for climbing without using the sharp leg irons?

Every pole has a little plate nailed on with an identifying number. Can you find it?

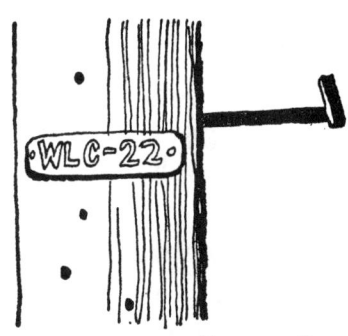

Since the wires sometimes carry high voltage, they must be insulated from the pole to prevent leakage of electricity. Can you see how each wire is supported on the pole by glass or other type insulators on the crossarm?

Observe how wires are covered by heavy protective sleeves of rubber or other insulation as they cross each other or go through the branches of a tree.

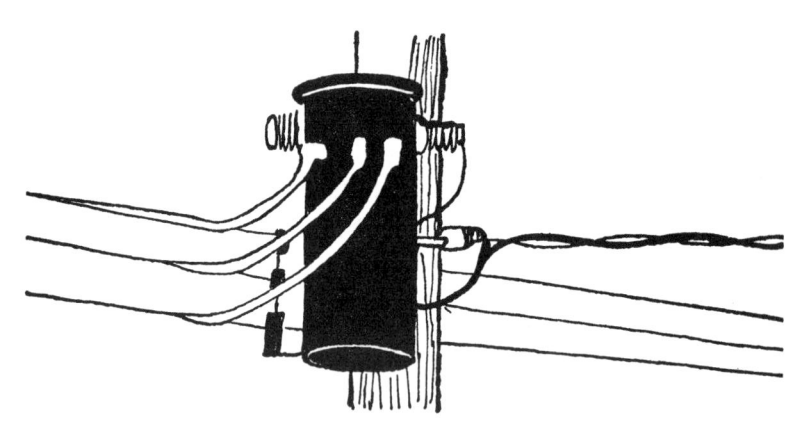

Those large black cans you see on poles are transformers. They change high voltages to low voltages. For example, 2300 volts may be changed to 115 volts. This safer voltage is then brought into a home. Look closely and you will see very heavy wires for high voltage entering the transformer and thinner wires coming out and leading to a house.

Do you think that if you walked far enough you would come to the power house or to another transformer? It should be interesting to track down the source of your house current.

If the weather is very warm, notice how the wires droop as they stretch between the poles. In the winter the same wires will be tighter. Linemen must allow for this expansion and contraction.

If you glance at the wires during a sleet storm you may be able to see how dangerously weighed down they become when covered with layers of ice. The electric company considers icing one of its biggest winter problems.

You may see a pole attached to a stranded steel wire which is anchored in the ground or attached to another pole. This supporting wire is called a guy wire. Observe the different ways it is used to relieve strain on a pole. Notice how the wire is attached to the pole. There is usually a wide piece of metal between the wire and the wooden pole. This spreads the force over a wider area and does not allow the wire to cut into the soft wood.

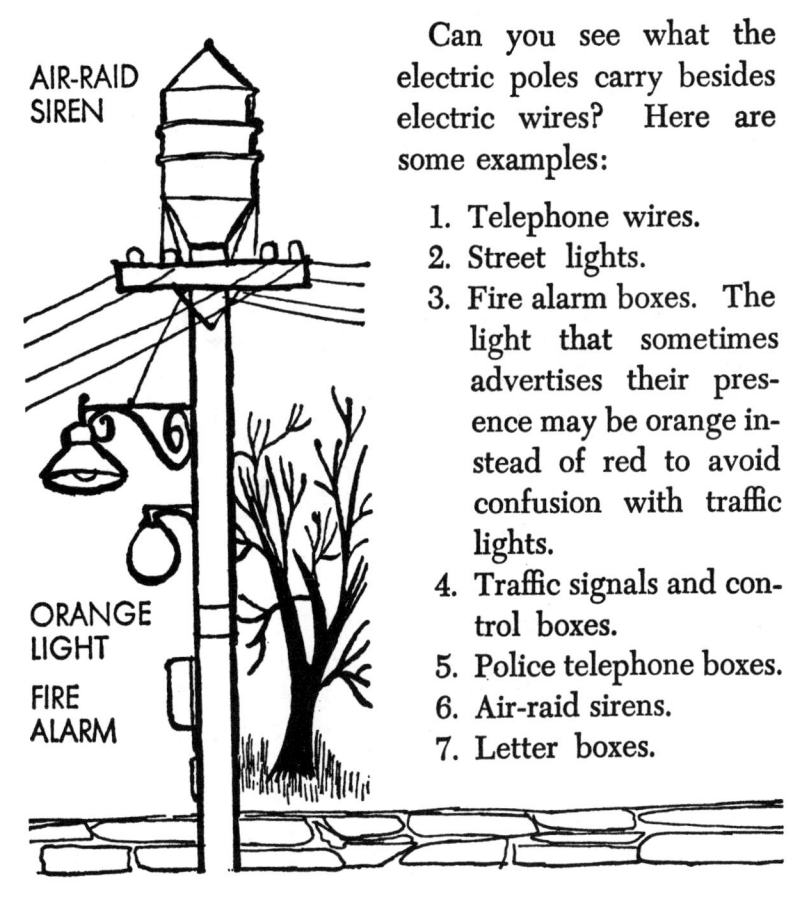

AIR-RAID SIREN

ORANGE LIGHT
FIRE ALARM

Can you see what the electric poles carry besides electric wires? Here are some examples:

1. Telephone wires.
2. Street lights.
3. Fire alarm boxes. The light that sometimes advertises their presence may be orange instead of red to avoid confusion with traffic lights.
4. Traffic signals and control boxes.
5. Police telephone boxes.
6. Air-raid sirens.
7. Letter boxes.

YOU CAN FIND OUT MANY THINGS FROM
THE INSCRIPTIONS ON THESE COVERS

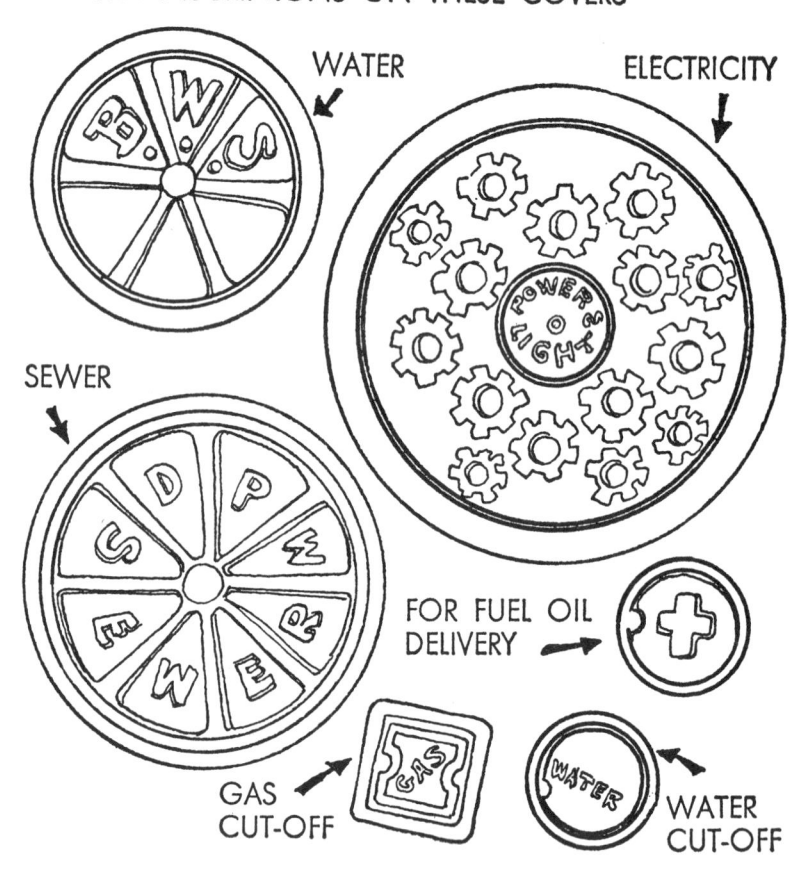

More to find out

1. Try to discover how certain streets received their names. Some may have been named for a natural feature, such as Oak Drive, Flatlands Avenue, North Street, Ridge Boulevard.

2. By reading the inscriptions on the iron manhole

covers in the street, find out which are for sewers, electricity, gas, telephones, and drinking water.

3. On the sidewalks find the screw cap for the fuel oil delivery pipe.

4. Can you find the iron covers for water and gas cut-off valves for a house? Look in the sidewalk near the curb.

5. Find out from the Highway Department what they use to keep down dust on dirt roads.

6. Sometimes the Highway Department uses sand to prevent skidding on icy roads. What other materials do they use?

7. Find out from the Sewage Department where the sewage in your neighborhood goes.

8. Ask several men why they walk on the curb side of a sidewalk when escorting a lady. This custom goes back hundreds of years.

9. By keeping records, find out whether your street lights go on every night at exactly the same time. In some communities the lights go on automatically when the sun sets.

10. Look for dark oil stains on concrete roads, especially on hills and curves. Cars work harder on hills, and drop more oil. On sharp curves, drops of oil under the engine are shaken loose.

11. Put your ear to the traffic signal control box which is found on some corners. Listen to the sounds of the electric clock timing device inside.

12. Ask any car driver what the dotted, single and double lines in the middle of the road mean.

LENDING AN EAR TO SOUNDS

No matter where or when you walk you are always completely surrounded by a great variety of noises. Strangely enough, most of the time you do not hear them unless you make a definite attempt to do so.

You have trained yourself from childhood not to listen to most of the customary sounds. But now, in order to make your scientific walking jaunts more exciting, you should try to analyze or track down some of these sounds. It can be fun to find the origin of a strange sound by making it into a game or treasure hunt. Call it a "sound-hunt."

If you stop for a few minutes and listen carefully to every sound, you will be surprised to find that in the "silence" around you there are many noises indeed. First and foremost today are the sounds of automobiles starting, stopping, popping, squealing, or hissing. Then there are airplane noises, bird calls, church bells chiming, and electric tools whining. You may even hear the rustling of leaves!

Before we go any further let us recall a few facts you should know about sounds. Every sound is caused by the rapid back-and-forth movement of a substance. These *vibrations,* as they are called, can be produced in a great many ways. You also learned that the greater the number of vibrations per second, the higher the pitch or tone.

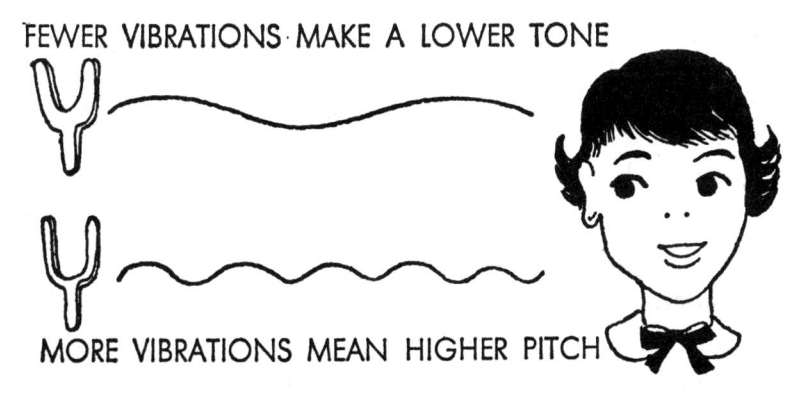

FEWER VIBRATIONS MAKE A LOWER TONE

MORE VIBRATIONS MEAN HIGHER PITCH

In addition to the pitch of a sound, there is also a certain quality possessed by a vibrating substance. It is this characteristic of a sound which enables you to tell the difference between a violin and a flute.

Since automobile noises are the most frequently heard sounds on the streets today, let us discuss some of these as an illustration of what you can do with other sounds.

One of the most familiar sounds is the starting of a car's engine. First there are a few low-pitched whines, caused by the slow turning of a powerful six- or twelve-volt electric motor which is geared to the gasoline engine. As soon as the engine starts, its fast motion disengages the starting motor. The sequence of all of these sounds tells an interesting story, if you know how to listen.

23

Direct your attention to an automobile that has stopped for a red light. You probably will not be able to hear the engine working because it is idling now. The motor is receiving just enough gasoline to prevent it from stalling. There is an adjusting screw on the carburetor (KAR-byoo-ret-er) which controls the flow of gasoline to the engine after the foot is removed from the accelerator (ak-SELL-er-ay-ter) pedal so that the engine is turning at its slowest speed. No wonder you do not hear it. There is another reason too, for directly under the hood may be cemented sound-insulating material of spun glass. The next time you see some cars in a service station with their hoods up, look for this.

Sometimes, however, an idling engine may be noisy because it is "racing"; that is, the speed is advanced too much. Or it may be that, because of fouled spark plugs or poor carburetor adjustments, the explosions do not occur in an orderly fashion and the engine "gallops," "rolls," or "motorboats," as the mechanics say. Your ear can very easily pick out such a poorly adjusted engine from many others which are waiting at the red light.

Some cars are particularly noisy because they may have defective mufflers. A muffler is a sound-deadening apparatus through which exhaust gases have to pass.

If the gases produced by the explosions in the cylinders were to escape directly into the open air, they would make very loud popping noises like a motorcycle. To prevent this, the gases are forced to go over a longer path, through baffles or compartments set up in the

EXHAUST GAS COMES
INTO THE MUFFLER
HERE

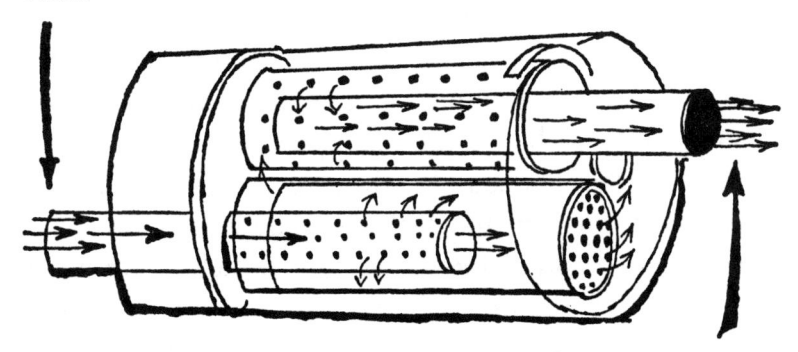

AND COMES OUT HERE

muffler. The walls of each chamber absorb a little energy of the vibrations. By the time the gases reach the end of the path, they have too little energy to make much impact against the outside air.

In a defective muffler the steel has been destroyed, mainly from the corrosive vapors and liquids in exhaust gases. When a hole develops, the gas escapes into the outside air before it is slowed up.

HOLE IN MUFFLER

Let us go back to the traffic light as it changes to green. As every motorist steps on the gas the engines get very noisy. You can hear things speeding up, loud exhaust sounds, and you are aware of much activity under the hoods. The reason is that automobiles have to use extra energy to get rolling from a stopped position.

Another familiar sound is the screeching of tires when a car makes a turn too fast. A vehicle making a fast turn tends to skid in the direction in which it was going before it made the turn. The tire rubs against the pavement and produces the squeal. The same sound is sometimes produced by your sneakers when you make a sudden stop on the gym floor.

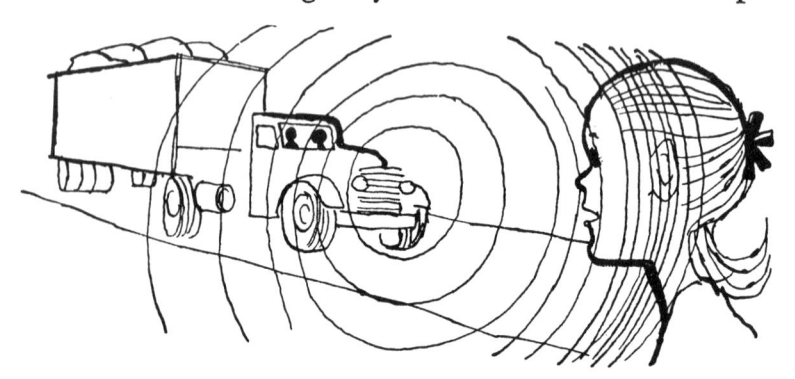

You may have wondered about the peculiar changing of tone of an engine as a truck approaches and then passes you. It is particularly noticeable when a horn is being sounded at the same time. Scientists call this "the Doppler effect." When you first hear the sound of a horn it has a certain pitch because of the number of vibrations per second it produces. As it approaches you, the effect is to bring to your ear more vibrations per second than if it were standing still. Therefore you hear a higher tone. As the fast-moving truck swishes by and away from you, the number of vibrations per second reaching you is less than before. Therefore the tone is lowered. Of course, the sound is louder too as the truck approaches. Do not confuse loudness and higher pitch.

In this modern world, airplane sounds have become commonplace. You can find an airplane overhead without looking, if you close your eyes, cup your hands over your ears, and turn your head toward the loudest sound. Open your eyes and you will be looking at the plane.

Before radar was developed this technique was taught to all soldiers.

Of course, jet planes may be moving so fast or so high that you will have to look *ahead* of the loudest sound. By the time the slow sound vibrations reach you, the fast jet is far out in front. Try it to be convinced.

With a little practice you should be able to tell the difference between a jet and a propellor plane without looking up.

Another familiar sound that your ear can pick out from thousands of others is the sound of a train going over the spaces between the tracks. The standard length of rail in the United States is 39 feet. Since expansion joints are left between rails, and each car usually has eight wheels which strike these joints, you can understand why there are so many clicks. The faster the train goes, the more frequently do you hear the clicks. With experience you can tell, just by listening, whether a train is going slowly, moderately, or very fast.

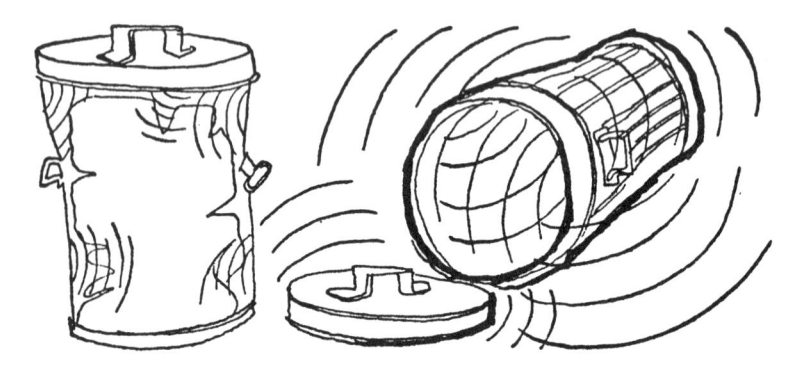

On one of your walks you may hear the sanitation men collecting the garbage. Why do the empty cans always seem to make such a clatter? There is an interesting reason for this. The sounds are caused mainly by the vibration of the metal can itself. When a can is full, this is not usually too loud because rubbish actually stops most of the metal from vibrating. When the can is empty there is more metal surface free to vibrate and, what is more important, the air inside the can vibrates too. This increase in loudness of a tone is called *resonance* (REZ-uh-nuhns). The vibrations which are strengthened are only those with which the air in the can is able to vibrate in unison. You get the same effect when you yell while walking through the tunnel of an overpass.

Have you ever wondered why you can hear the rustling of leaves on a windy day? You may not be able to hear a few leaves scraping against each other, but when many thousands do the same thing a whisper becomes a shout!

These examples of neighborhood sounds should give you some idea of how much scientific fun it can be if you *stop,* look, and LISTEN!

More to find out

1. Learn to distinguish the sounds of a helicopter, one-, two-, and four-engine airplanes.

2. Ask a mechanic what causes an automobile to backfire.

3. Why do you hear the hissing sound of escaping air when a bus or large truck stops?

4. Why does the little ball in a police whistle help produce such a penetrating sound?

5. Ask a fuel oil delivery man why there is a musical hum while the oil is being pumped into the tank in a house.

6. Are there any sounds in your neighborhood which occur every day at the same time?

7. At this moment, how many neighborhood sounds can you hear and explain?

8. What is the reason for the whistling of the wind?

9. What is the most distant sound you can hear while standing in front of your home?

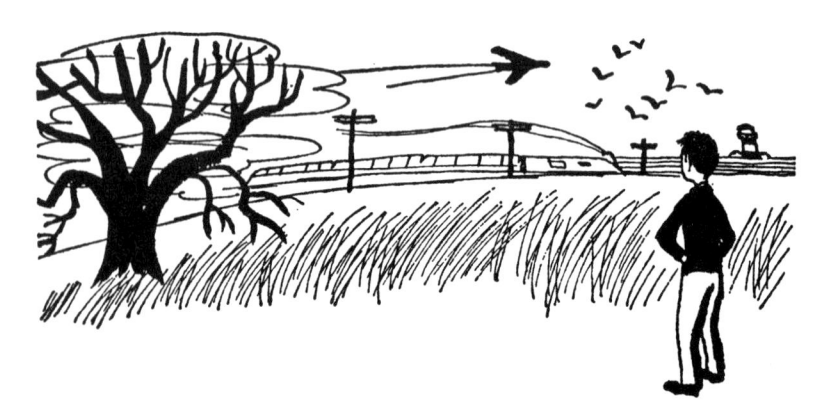

BECOMING AWARE OF ODORS

You can get accustomed to odors just as you do to sounds. Unless you make a conscious effort to become aware of the dozens of enjoyable, as well as the unattractive smells, you may easily walk right past them.

You can educate your extremely sensitive nose to recognize different odors when you pass stores and factories or when you go through the fields and woods. Try to discover where the smell comes from and what caused it. This should add to your knowledge and make your walks more pleasurable.

An odor is caused by gas or tiny particles suspended in the air. Since molecules are in constant motion, they keep spreading away from their source, even in still air. Of course, a wind carries these particles faster.

You usually have to sniff deeply when you wish to smell something really well. This is because you have to get the odor into the upper rear part of your nose. That is where you have the tiny nerve endings imbedded in the moist membranes. From there the nerves go up to the smelling center in the brain.

WHICH DO YOU SMELL MORE THAN YOU TASTE?

The sense of taste is closely connected with smell. Sometimes when you think you are smelling certain flavors or spices you are probably tasting them.

If you have an allergy or a cold you may lose your sense of smell for a while because your nose membranes have become inflamed. There are some people, too, who unfortunately have lost the ability to smell things.

Another interesting fact is that you get used to an odor after you smell it for a while. Even if it is a bad one, you become unaware of it.

Let us discuss some of the odors you are most likely to encounter on your walks.

Automobiles are responsible for many of the odors in the streets today. Indeed, many scientists believe that they are the biggest cause of the irritating smog which lingers in so many large cities. Smog, as you probably know, is a mixture of smoke and fog.

Theoretically, if the gasoline were completely burned in the cylinders of an automobile engine, carbon dioxide and water vapor would be formed. These are two odorless, harmless substances.

But somehow, the burning is not complete and the exhaust gas of a car contains some smoky, smelly, unburned gasoline and oil products, and dangerous carbon monoxide.

Carbon monoxide is a deadly invisible gas which has no odor, and it is for this reason that you should never stay in a closed garage with the automobile engine running. In fact, do not have the car windows entirely closed at any time that the engine is working. A leaky muffler may send carbon monoxide up into the car.

When you walk past a car you may also smell the

GASOLINE FROM THE GAS TANK

HOT OIL FROM THE CRANKCASE

ALCOHOL FROM THE RADIATOR

fumes of very hot oil. Most of the moving parts of an engine are in the crankcase, which contains about five quarts of oil. There is a breather pipe underneath the engine from which the hot fumes escape. In addition, the surface of an engine is rarely clean, and when it gets hot the oil on the metal starts smoking. This is what gives a car its characteristic odor.

You can also smell gasoline as it evaporates from the gas tank, because its cover must be partly open to the outside air pressure. If a car is kept in the hot summer sun, much gasoline can be lost this way.

In the wintertime, especially on a warm day, you may also smell alcohol as it evaporates from the radiators of some cars.

Still another odor is that of burning rubber from a car making a panic stop. The motorist loses many weeks of tire wear in the few seconds when he does this.

Flower odors are highly enjoyable whenever they cross your path. It is pleasant to recognize the fragrance of lilacs, roses, hyacinths, honeysuckle, and scores of other scents. Learn to associate the odor and the sight of the plant. You can become very good at this.

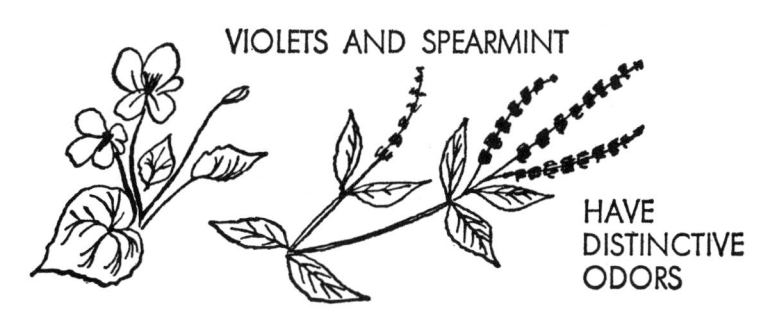

VIOLETS AND SPEARMINT HAVE DISTINCTIVE ODORS

Who can walk by a linden tree in bloom and not want to know what is the source of the overpowering sweet odor? New-mown hay, too, has an unforgettable aroma. So has pine.

The sweet odor of flowers attracts insects and animals which carry pollen from one flower to another. The fragrance comes from the nectar and tiny sacs in the petals and leaves.

As you brush against the leaves of sassafras, bayberry, or mint, crush a leaf or two to get the full odor.

Now at the other extreme are the distinctive odors of garbage, sewage, or decaying dead animals. There is a chemical reason for these putrid smells. The flesh of all living things is composed mainly of protein, a complicated substance consisting of such elements as sulfur, phosphorus, carbon, oxygen, hydrogen, and nitrogen.

When bacterial decay or chemical breakdown occurs there is a reshuffling of these elements into simpler compounds. It happens that some of the new substances formed are gases which have particularly foul odors. For example, hydrogen and sulfur form hydrogen sulfide, which has a rotten-egg odor.

But some bad odors have important uses. Have you ever passed an excavation in the street where the gas company is repairing a gas pipe? Did you wonder about the very bad smell that came from the ditch? Cooking gas contains carbon monoxide, a poisonous, odorless gas. In order to warn of its presence, the company introduces into its gas supply a substance which has such a bad odor that people at home cannot disregard it.

In some industrial sections, soft (bituminous) coal is still used. Coal-burning locomotives also use this coal. The dense smoke contains smelly hydrogen sulfide, among other things. Many cities have laws which do not allow soft coal to be used. Industries spend millions of dollars for research on smoke control.

If there is any fresh paint nearby, you can detect the well-known odor of turpentine and benzine. These are used to thin the oil and the pigment in the paint. Sometimes when you pass an auto-body repair shop, you can smell lacquer thinner, which has the odor of nail polish.

After you get to know the smells in your neighborhood you may become an expert in detecting the direction of the wind just by using your nose. If the wind comes from the ocean side of town, then there is that salty, even fishy tang to the air. Wind from another side may bring a slight odor from the town's incinerator or dump. The wind has a distinctive odor when it blows from marshes, mills, bakeries, chlorinated swimming pools, slaughter houses, pine woods, or perhaps a chocolate factory.

Have you ever taken a walk after a rain has washed down most of the dust that was in the air? There is a clean smell you cannot easily forget!

When you are on a smelling expedition, see how other animals use their sense of smell. A dog, for instance, has such a keen sense of smell that its brain center for smelling is very much larger than ours. Dogs find out a great deal about the world around them by using their noses.

You may meet a harmless garter snake. See how its forked tongue darts in and out of its mouth. It uses the tongue to bring odors into its smelling organ in the roof of its mouth.

Certain insects smell with their antennae. How in the world did scientists ever find that out? Very simply! They discovered that when they cut off the antennae the insect could not find food which was only a foot away!

Try to find the most prominent odor in certain stores, offices, and other places. (See the chart for suggestions.)

Place	Most common odors
Grocery	old milk smell, cheeses
Delicatessen	vinegar, spices
Shoe-repair shop	shoe polish, benzol used in rubber cement
Barber shop	talcum powder perfume, bay rum, alcohol
Laundry	chlorine bleaches, perfumes in soaps
Bakery	yeast, vanilla, cinnamon
Restaurant	coffee, frying bacon, tobacco
Dry cleaner	naphtha, carbon tetrachloride
Newsstand	printer's ink
Fruit store	bananas, apples, onions
Dentist	oil of cloves
Hospital, doctor	disinfectants, alcohol
Swimming pool	chlorine
Gymnasium	rubber sneakers
Clothing store	chemicals used in mothproofing
Paint store	turpentine, benzine, linseed oil
Auto-body repair shop	lacquer thinner
Furniture store	chemicals in the polish
Lumber yard	pine, cedar

More to find out

1. Open a bottle of perfume in one corner of a room where the air is quiet. Sit in an opposite corner. How long does it take the odor to reach you?

2. Close your eyes and try to identify the substances placed under your nose by a friend.

3. When you smell a new odor anywhere try to track it down by asking the person who might know, or by your own detective work.

4. Stay in a place where there is a strong odor. At certain intervals find out if you can still smell it.

5. Learn to recognize ammonia, kerosene, alcohol, turpentine, benzine, carbon tetrachloride, laundry bleach (chlorine), vinegar, lacquer thinner, other solvents.

6. Learn to recognize different spices. Try cloves, bay leaves, cinnamon, nutmeg, paprica, dill, ginger, caraway, and others your mother may have.

7. When walking with a friend, find out who can smell more known odors in a certain place.

8. Why do marshes sometimes have a bad odor?

9. Have a friend close his eyes and hold his nose. Place a slice of apple in his mouth. He will probably not be able to identify it. Try a slice of pear next. Does this show that smell and taste are sometimes related?

RECOGNIZING BUILDING STONES

An intriguing thought to dwell upon when you are out for a walk is that mankind is still using some of the building materials used by cave men before the dawn of history. The Incas, Egyptians, and cliff dwellers of Arizona built roads, monuments, and homes from the same stones being used today for modern homes.

It is easy to learn to recognize many of the stones used in construction. You will discover that there are about a dozen varieties which are used. Learning to identify them is like remembering the faces of your friends. After a short while, you will be able to distinguish granite, marble, slate, limestone, sandstone, and other stones.

Do not be discouraged if you cannot name every rock you find in the fields. Even geologists have difficulty doing that.

There are many ways of identifying unknown rocks. One of the best ways is to ask people who can give you the correct answers. Friends who have studied geology or those who collect rocks may be able to help you. Ask your science teacher. Visit a museum. Compare your stone with others in a labeled rock collection.

Very frequently the owner of a house may know the stones used in his building.

Granite (GRAN-it)

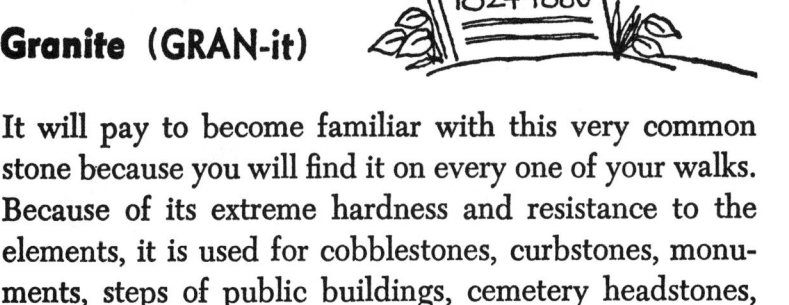

It will pay to become familiar with this very common stone because you will find it on every one of your walks. Because of its extreme hardness and resistance to the elements, it is used for cobblestones, curbstones, monuments, steps of public buildings, cemetery headstones, piers of bridges, and ornamental fronts of banks, libraries, schools, and other public buildings.

There are many varieties of this useful stone, as its crystals may be very fine or extremely coarse. Granite is classified as an *igneous* (IG-nee-us) rock, since it was formed from the hot molten state deep in the earth. Extreme pressure or volcanic action brought it to the surface.

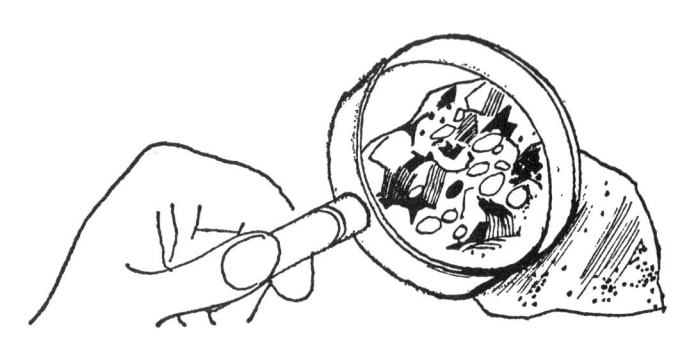

Granite may be given a very high polish to bring out its beauty. It usually comes in gray tones, but there are also interesting pink varieties. It may be recognized by its speckled appearance, which comes from shiny flakes of mica, hard glossy quartz crystals, creamy white or pinkish feldspar, and flecks of a greenish-black substance called hornblend. Use a hand lens to examine all the crystals closely.

Marble

Of course, you know this beautiful stone because of its interesting designs. It takes a glasslike polish, and comes in white, black, and many shades of pink and green.

It is used to decorate the interior and exterior of buildings. Vestibules of apartment houses and public buildings frequently have marble walls. Banks may even use it for outside walls.

Marble was formed from the chalklike skeletons of sea animals that lived millions of years ago. When these animals died they formed a sediment which the great pressure of the ocean cemented into a limestone. Further pressure and heat through millions of years changed this into marble. Geologists call marble a *metamorphic* (met-uh-MOR-fik) rock, since it was changed from limestone to its present form.

The trouble with marble is that it is not durable, and is easily scratched. Steps of buildings will show signs of wear after a few years of use. You can test this for yourself by placing a straight edge of a long board or ruler across the step. If the step is worn, you will be able to see light between the step and the board.

Another difficulty with marble is its reaction to acids. A powerful acid like hydrochloric acid will act on this stone, making it disintegrate very fast and give off bubbles of carbon dioxide. And even vinegar, which is a weak solution of acetic acid, will also cause this chemical reaction. Try it on an old piece of marble.

Do you know that rain water also eats away marble? Rain is a very weak acid because as it falls it dissolves some of the carbon dioxide from the air and forms carbonic acid. Any marble or limestone which is out in the rain will be very slowly eaten away.

Because of this chemical erosion, the ancient marble statues and parts of buildings we find in Greece today are no longer as sharply chiseled as they once were. Cemetery tombstones made of marble become difficult to read after many years. As a result, modern tombstones are often made of granite.

Limestone

This natural stone is widely used today because it is not as expensive as other building stones. It is also relatively easy to shape.

At first sight it looks like cement. It has a fine-grained, gray or creamy appearance. If you look closely, however, you may see parallel ridges all over it, made by the chisels which cut it. You may even see fossils of ancient animals and plants imbedded in it. It can be easily scratched with a knife. Like marble, it consists of calcium carbonate so that vinegar and other acids will act on it.

It is used extensively for window sills and the fronts of public buildings. Large office buildings are often made entirely of limestone.

The pyramids of Egypt were built of limestone blocks.

Slate

You need no introduction to this familiar classroom-blackboard stone, which may also be brown, purple, or green. Although it is expensive, it is often used for roof shingles, for it lasts a lifetime. You will see slabs of slate being sold by a local gardener for flagstone walks, and in very old neighborhoods you may still see slate sidewalks. This is not a good use for this stone because it is brittle and scales easily.

Slate, like marble, is a metamorphic rock, having been changed by heat and pressure from shale, a *sedimentary* (sed-i-MENT-a-ree) rock. Slate is cut up into chunks in quarries and then split into sheets by means of thin, flat chisels.

Shale

This is often called mud-rock. It is a sedimentary rock, formed when clay and mud were deposited in water and hardened by pressure. When wet, it actually smells like mud or clay. A knife scratches it easily. There are some shales so soft that they crumble in your hand;

others are very much harder. That is why you may have some trouble identifying them.

Shale can be gray, black, or red. Because shale can be split into sheets, it is frequently used for sidewalks.

Sandstone

In some towns or cities, many houses can be seen with this type of construction stone. Brownstone houses used to be very popular, especially when this reddish-brown sandstone could be obtained locally.

Look closely and you will notice the scaling which occurs on some very old houses. This is caused by expansion and contraction due to temperature. The weather is not very kind to brownstone walls.

Sandstone, which is a sedimentary stone, consists of hard grains of sand loosely cemented together. If a piece is broken you can feel these sharp grains of quartz as you run your finger over the edges. Scrape some grains from the surface with a coin. See how the coin becomes scratched.

Bluestone

This is a stone very commonly used for sidewalks, window sills, and curbstones. It is a sandstone, but is frequently mistaken for a hard shale.

Basalt (buh-SAWLT)

This is probably the heaviest stone you will find, because it contains a considerable amount of iron; it is also one of the darkest common stones. It is used for cobblestones and curbstones and is also crushed and used for roads. It is an igneous rock, very hard and durable.

Gneiss (nice)

This stone has layers of fine minerals. These bands are usually twisted and bent. It is a durable metamorphic building stone.

Mica schist (MY-kuh shist)

This is a metamorphic rock, having been changed from impure shale. It is very easy to identify. Look for pale amber or black layers of shiny mica, which can be picked out without too much trouble.

Sand

This consists mainly of tiny grains of worn-down quartz and other rocks. Use a magnifying glass to see the glass-like hard quartz, as well as red and black minerals. Sand is a hard substance. When cemented together in nature it forms sandstone. Sand is used in mixing concrete.

Rust stains in stones

This shows the presence of some iron compound.

Coal

This is actually a rock that burns. The most greatly changed form is hard coal, called *anthracite* (AN-thruh-site). It was changed from soft *bituminous* (bi-TYOO-mi-nus) coal.

Asbestos* (as-BES-tos)

The fibers from asbestos rock are made into many kinds of roof shingles you may see. The big feature of this material is that it is fireproof.

*[Publisher's Note, 1991:] Many years after this book was written, asbestos was discovered to be unsafe. Most of it has now been removed from buildings. If you should find any, don't handle it!

More to find out

1. Bricks are made by baking a mixture of clay and sand. Watch bricklayers at work, mixing mortar, using levels and plumb lines. Are all bricks alike?

2. Tile is made like bricks, but a glaze is baked on the surface. Look for it on chimneys, roofs, sewer pipes, electric-cable conduits, and colorful store fronts.

3. Look for cement and cinder blocks. See where these are used for building foundations, small garages, and walls. Cinder blocks are darker than cement blocks and have a coarser texture.

4. Find out how glass is made by consulting an encyclopedia or library book. See it used for colored store fronts, insulators on poles, and in many other places.

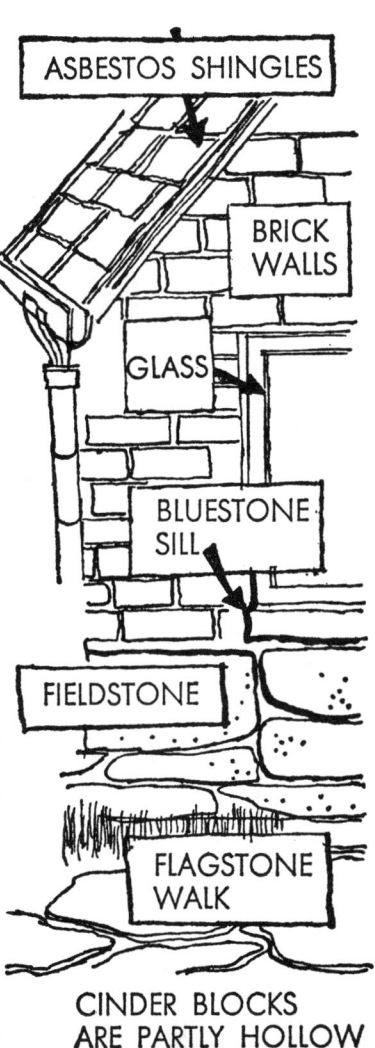

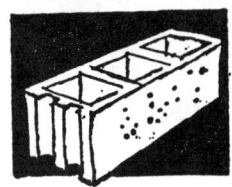

5. Make a labeled collection of many different stones that you find.

6. Make a chart showing where different stones are being used on buildings in your neighborhood.

7. Visit a mason's supply yard to see natural and artificial stones.

8. Separate pieces of mica from a rock. Can mica conduct electricity?

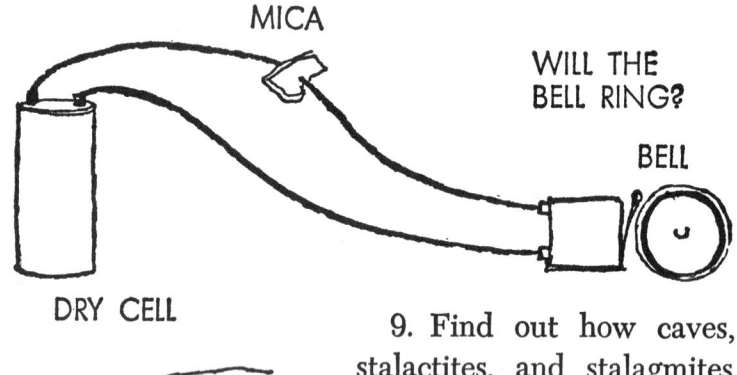

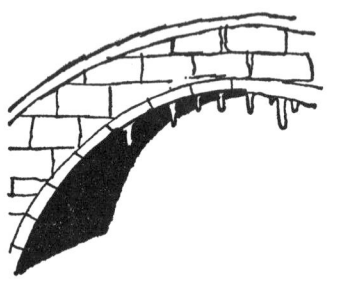

MICA

WILL THE BELL RING?

BELL

DRY CELL

9. Find out how caves, stalactites, and stalagmites are formed. Look for small stalactites hanging from concrete archways.

10. Ask a hardware dealer how signs and brackets are fastened to stone.

METALS SEEN IN THE NEIGHBORHOOD

Did you know that almost all the metals that you see on your walks were once part of rocks called *ores,* and had to undergo long chemical treatments to make them useful?

Most metals cannot be used in the pure state. Either they are too soft or they are likely to have other objectionable features. Therefore, they are melted together with other metals to form mixtures of definite composition called *alloys* (AL-oiz), which can be used.

Becoming familiar with common metals is not as difficult as it seems. You will find that if you can recognize about ten of them, you already have a very good start. These include iron and steel, aluminum, copper, brass, galvanized iron, chromium-plated metals, lead, and several others.

As a young scientist you can observe how each metal

is being used for a specific purpose. After a while your sharp eyes will be able to detect where a metal is used incorrectly. You may also discover how a metal has been treated to prevent it from rusting or tarnishing.

Metals are used mainly because they are strong and hard. When properly protected, they withstand the elements much better than wood. They are also fireproof. They can be made into beautiful shapes and colors and do not usually require much care. And of course, there may be other special reasons for using certain metals.

Iron and steel

One of the most common metals is iron. It is cheap and strong and can be shaped in many ways.

Its one disadvantage is that it rusts easily. Look around you for the telltale signs of rust. You will find them everywhere you turn. Rust is formed when iron combines with oxygen in the presence of water. It can usually be prevented by painting the iron. This provides a waterproof membrane over the surface. You have probably seen iron painted with a special orange-red paint before the final color is applied. This helps prevent rusting.

Sometimes iron is dipped into molten zinc. A layer of zinc forms over the iron and helps protect it from rusting. The product is known as galvanized iron. Your garbage can is probably the best example. Sometimes the zinc has a crystalline appearance.

Iron is magnetic even when it is covered. See how a compass needle is attracted by some iron object which is covered by paint or plating.

Iron is obtained from iron ore, which has many impurities in it. The ore is mixed with coke and limestone and emptied into tall blast furnaces. Because coke is porous, very hot air can be blown through the mixture in the furnace. The coke, which is mainly carbon, combines with the oxygen in the iron to form carbon dioxide, which escapes. The limestone removes the other impurities, and the heavy molten iron sinks to the bottom. When clay plugs are opened, the iron pours out and is cast into molds called *pigs*.

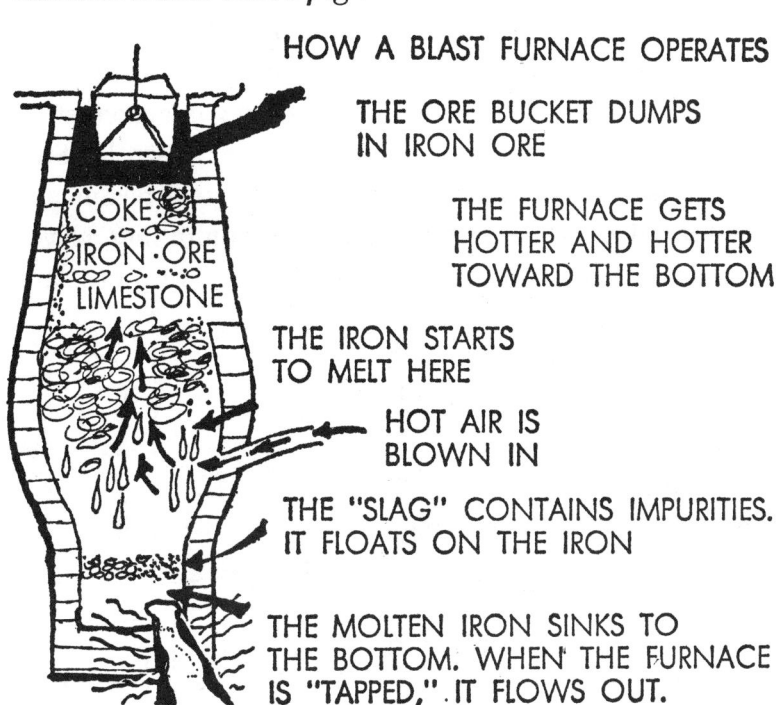

HOW A BLAST FURNACE OPERATES

THE ORE BUCKET DUMPS IN IRON ORE

THE FURNACE GETS HOTTER AND HOTTER TOWARD THE BOTTOM

THE IRON STARTS TO MELT HERE

HOT AIR IS BLOWN IN

THE "SLAG" CONTAINS IMPURITIES. IT FLOATS ON THE IRON

THE MOLTEN IRON SINKS TO THE BOTTOM. WHEN THE FURNACE IS "TAPPED," IT FLOWS OUT.

This cast iron, or *pig iron,* as it is called, is quite brittle and is used for cheap castings and in places where it is not subject to banging or bending.

Most cast iron however, is used to make steel. Steel is purified iron which contains less than 2 per cent carbon. By varying the amounts of carbon, by the addition of certain metals, and by various heat treatments, it is possible to get special steel alloys.

For example, stainless steel contains chromium. This hard, rustproof steel is used for store fronts. Steel used for automobile springs may contain vanadium. Hard steel tools are made by adding tungsten. Steel safes contain manganese.

Almost all the iron objects you may see on your walks are really steel alloys. Even a common nail is made of steel. Galvanized iron should be called galvanized steel.

Do you know that houses use many tons of steel? Stop at a house being built and observe the steel beams and the long stretches of iron pipes used for plumbing. Even the sinks and bathtubs have steel inside the porcelain.

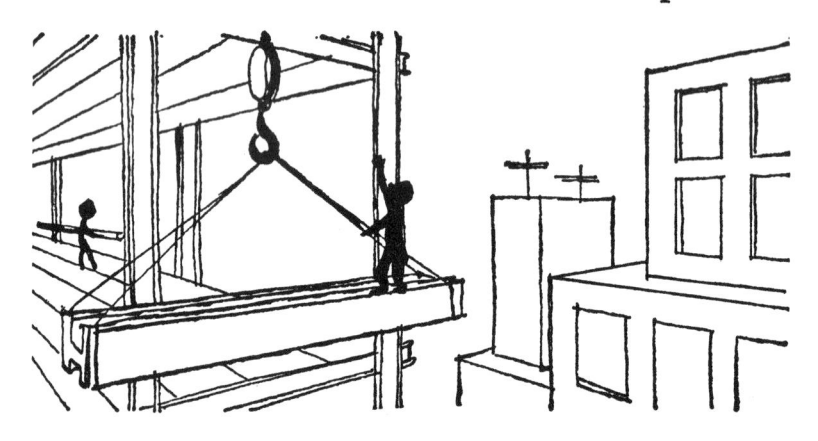

Notice the long iron sewer pipes being installed under the streets in a new neighborhood. See how they are covered with heavy asphaltum paint to retard rusting. The automobiles you see on the road are mainly made of steel. Perhaps you may see a mechanic welding the steel at a body repair shop.

Aluminum (al-LEW-min-um)

It is hard to believe that this common, relatively cheap metal was so expensive about fifty years ago. Although it is the most abundant metallic element, it is found in ores from which it is extremely difficult to extract the aluminum except by an electrical process.

The biggest advantage of aluminum is its lightness. The airplane overhead uses aluminum extensively. The pure metal is quite soft, but it has been combined with many metals to produce hard alloys which are very useful.

Since it does not rust like iron and needs no painting, it is used everywhere. As a matter of fact, aluminum does form a thin, transparent coating on its surface, but this only protects it against further corrosion.

The silvery, metallic doors, window frames, screens, and roofing pipes on buildings are aluminum. So are television aerials. You will recognize this distinctive metal in store fronts and in the ornamental facings of many buildings. Lightweight beach and lawn furniture is constructed of aluminum. You will also see ladders made of this metal. Look at the trim on buses and other cars. They may be aluminum instead of chromium-plated iron. Sheets of aluminum foil are used as insulating material in the walls of new buildings because they reflect the heat back into the house in the winter, thus preventing its escape. In the summer, aluminum cuts down the entrance of heat in the same way.

ALUMINUM INSULATION

Examine a discarded candy or cigarette wrapper. People may call it tinfoil but it is probably aluminum. This foil is also used in the kitchen for wrapping food.

Aluminum paint consists of fine aluminum powder held in a varnishlike liquid. It forms a durable paint and reflects sunlight. Bridges and huge gasoline and oil-storage tanks are painted with it.

Copper

You should have no trouble recognizing this colored metal because it looks like a penny. This metal was one of the first to be used by primitive man when he discovered pure veins of it.

Copper is quite expensive, and for this reason, it is not used more. Its big advantage is that it does not rust. It will tarnish, but this film only protects the surface from further damage. You have seen bright copperwork turn brown and finally black. On your walks you will also see the greenish discoloration on the copper roofs and steeples of churches and public buildings. Statues in the park made of brass or bronze, which are copper alloys, also show this greenish stain.

GREEN DISCOLORATION

The way to prevent copper from tarnishing is to shine it up first and then brush or spray on a protective film of clear lacquer.

Copper is used for window screens, roofing, and plumbing—and around ships, since it resists marine growths.

Copper is an excellent conductor of electricity, and almost all electric cables you see use copper wire. The glittering gold paint used ornamentally is made up of 90 per cent copper powder with aluminum added. The red lenses used for traffic signals have a copper compound in the glass.

57

Lead

This heavy, dull-looking, gray metal can be seen on all of your walks if you know where to look for it.

The telephone cable on electric poles may be covered by this soft, inactive metal. A thickened part of the cable shows where splicing was done. Heavy electric cables going underground are often sheathed with lead. You may see a pot of molten lead near an open manhole where men from the electric company are working. The lead is used to cover splices and to make them watertight.

Lead is resistant to weather and also to most chemicals. It is easy to bend, cut, and solder.

Lead pipe is also used to make flexible, long-lasting connections in the sewage pipes in a home.

Look closely where metal fences and railings are anchored into concrete. You may find that lead has been poured into the hole and hammered down tightly. This holds the fence firmly in place.

Lead is also used in storage batteries of automobiles. Watch a repairman at a gasoline service station grunt as he lifts a heavy battery that is filled with lead plates.

Tin

It is extremely doubtful that you will see a pure sheet of tin, and even the "tin can" used for food is only coated with a thin layer of tin. About 98 per cent of a tin can is steel.

Thin sheet steel covered with tin is called *tin plate*. It is used in sheet-metal work and is made into ducts for air conditioners and hot-air pipes. Its shiny appearance, heavier weight, and magnetic properties distinguish it from aluminum, which often looks like it.

59

Brass

This alloy of copper and zinc looks almost like gold. It does not rust like iron, but it does tarnish unless it is polished regularly or lacquered.

It is used in plumbing for pipes and fixtures, for hinges, door knobs, locks, hose nozzles, screws, ornamental work, ship fittings, and musical band instruments.

Sometimes an object looks like solid brass but may be iron which is brass plated or painted to resemble brass. To test whether an object has iron underneath the brass surface, hold a compass needle near it. If there is any iron, the needle will be deflected by it.

Chromium plating

The bumpers and trim on most automobiles are plated with hard, durable, nonrusting chromium, which is a metallic element. Store-front ornaments, hardware like door handles, hinges, and screws, have the typical mirror-like gleam of chromium.

More to find out

1. Visit a sheet-metal shop to see how metals are cut, bent over, drilled, welded, and riveted.

2. Make a list of the different metals used in your neighborhood. Give the reason why a certain metal was selected for a specific purpose.

3. Look closely at the gold-leaf signs on store windows. For a few cents you can buy one small letter in an art supply store. The letters are pure gold, one quarter of a millionth of an inch thick!

4. Mercury is a liquid metal at ordinary temperatures. See it in thermometers. Also, look closely at the sides of fluorescent lights. Those little silvery beads are mercury.

5. Solder is a low-melting alloy made of tin and lead. Have someone teach you how to join metals by soldering.

6. Make a labeled collection of many different metals.

61

NOTHING LASTS FOREVER

This is a changing world. Nothing remains the same forever. Some changes occur so slowly that it takes thousands of years before they are noticeable; others take place rapidly and may be destructive as well as constructive.

Just ask any discouraged homeowner about the peeling paint, rusting iron, leaking roofs, cracked sidewalks and warped windows. He will also remind you of the weeds in the garden, the rotting of wood, and the worn-out plumbing parts.

On your walks, look for evidences of this widespread breakdown that civilized man keeps battling. At the same time, you will observe how nature builds as it destroys. When soil is being washed away in one place, it piles up in another. When iron rusts or wood decays, certain minerals are returned to the soil.

The very soil we need and cherish so much was formed when rock was cracked into small pieces by ice, or sandblasted by the wind. When plants and animals died their bodies also became part of the soil.

So, while you are walking, look upon nature's destructive side philosophically. It is all part of the changing pattern of the world.

The suggestions offered in the following pages do not deal with the dramatic havoc wrought by tornadoes, hurricanes, and floods, but with the apparently casual damage that we take for granted. They are things you will see on a walk in a residential neighborhood.

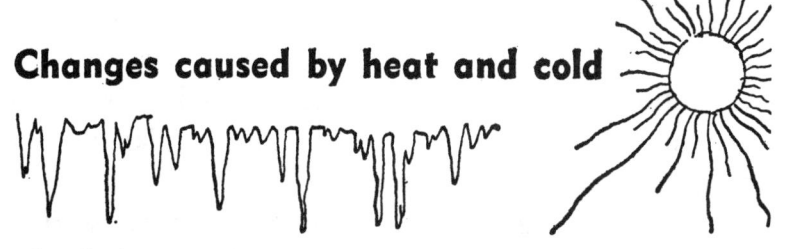

Changes caused by heat and cold

The daily, as well as the seasonal changes in temperature make things expand and contract all the time. This important scientific principle will help explain much of the damage to be seen on walks.

Cement sidewalks, concrete and asphalt roads develop cracks and bulges despite all the care taken during their construction. Not only does the road material expand and contract but the ground underneath it does the same.

That is why heavy statues in parks and cemeteries may be tilted, even though they have deep, heavy bases.

The walls of old wooden homes, garages, or shacks often appear not quite vertical because the ground beneath has risen.

Large rocks are often split or scaled when they expand in the heat of the day and contract during the cool evenings. Rocks and concrete roads will also split in the wintertime when water enters small cracks and freezes. Water expands as it changes to ice and exerts the tremendous force which does the damage. The condition of the city streets after a hard winter is proof enough of this.

As you pass some homes or stores look at the wood or metal window frames. They too expand and contract continuously. They may do so at a different rate than the walls of the house. This causes cracks, through which the heat of the house can escape and the harmful rain can enter. To prevent this, a flexible caulking putty is carefully applied at the open seams.

After several years the sun dries this material so that it can no longer stretch. Then you will see that openings have developed and new caulking has to be applied.

Rain and water produce changes too

You will not have to go to the Grand Canyon or to the Dust Bowl to see signs of erosion. Look around you and convince yourself that it is only a matter of time before everything gets washed into the sea—unless we do something about it!

Look in the gutter. Do you see some earth and sand? Try to discover where it comes from by trailing a telltale path. You may find that it came from your own lawn or from some place farther up the street.

Look in the gutter at the bottom of a street which is on a hill. You will find a big pile of soil and debris. It may even be piled higher than the curbstone. The earth and pebbles were deposited there by moving water which slowed down because of the decrease in slope.

This is the way rivers form deltas. As they near the sea and slow up they deposit the material that they were carrying along.

Look in the fields for a gully. Unless remedied, it will get larger and larger. Try to discover how it was formed. Could it be because there was no vegetation to hold the

65

soil? Did it start as a footpath which had prevented grass from growing? Or was it caused by the rapid flow of water which washed everything in its path down the hill?

Dig a little ditch, or better still, look at the sides of a fresh excavation. Notice the layer of darker earth near the top. It may be a few inches or more than a foot deep near the top. Feel how easily it crumbles between your fingers. This is the precious topsoil in which everything grows. It keeps man and animals alive. If it is lost by water or wind erosion it cannot be replaced.

It takes about 400 years to make one inch of topsoil from rock and decayed animal and vegetable matter. It is our country's priceless possession.

You have seen how it is being lost on a very small scale. All over this country millions of tons are being washed into the ocean by floods and dust storms whose damage might have been avoided. Our nation has already lost about one-third of its topsoil.

On your walks through the fields, see how certain sections having sharp slopes have become bare of vegetation because the topsoil has been washed away, leaving the nonfertile subsoil. The greater the slope, the greater the erosion.

People who purchase new homes and wish to have good gardens or lawns often must buy their own topsoil. You may see someone receiving a load of this expensive earth.

Sometimes there is an uncovered area between the sidewalk and the curb. The soil in that place may be washed away. See if the level is below the sidewalk or curb. The chances are that if there is grass growing there you will find less erosion than at a bare spot, for soil can be held in place by the roots of plants.

How do people prevent erosion when it rains or when they water their lawns and gardens? They have probably discovered that the best way is to keep the ground planted. If the lawn slopes very sharply they may have a series of level terraces, which allows the water to soak in slowly instead of rushing headlong to the street.

Look at the bases of trees. Are the large roots becoming visible, showing loss of soil? Fast-moving water will undermine trees. On a side of a hill, can you see the exposed fine roots of a tree? Can you suggest a way to save a tree which is being killed this way?

Examine the side of a building which has bare earth at its side. You will be surprised how high the earth has splashed on it because of the rain. Now look at a wall that has grass growing up to the building. You will find that the rain erosion is much less.

Look under small, flat, undisturbed stones. The earth right under the stone is higher than the surrounding earth, since it was protected from the splashing force of the raindrops.

After a rain see how quickly water soaks into topsoil, or soil where plants grow well. Water which collects in long-lasting puddles is probably in a hard-packed clay soil.

Whenever you see a fresh puddle, notice the muddy appearance of the water. After a while the material settles. This is the way harbors are filled with sediment. You can sometimes see dredges at work deepening channels for boats. Every few years it is necessary to remove quantities of this sediment.

When it rains, some water may get into the edges of the wooden windows and front doors. This makes it difficult to open them, because the wood cells get swollen as they absorb water.

You may also be able to notice other signs of warping.

Wind erosion

If soil is very dry, a strong wind can blow it away, especially if there is no vegetation growing on it.

On your walks look into some corner spot into which the wind is blowing. You may find a small pile of soil and bits of vegetation deposited there.

A solid fence will usually show a higher ground level on the side into which the wind blows.

In certain areas wooden snow fences or living fences (dense shrubs) are used to stop this loss of blown soil. Of course, the best method is to keep vegetation growing in the soil.

Chemical wear and tear

Everywhere you walk there will be iron objects rusting away. The reddish-brown stains are under iron fences, on buildings, on automobiles, on objects using iron nails, screws, or other hardware.

Millions of dollars worth of damage are caused by rusting—truly one of the destructive sides of Nature.

Another chemical effect is the fading of paints on houses and signs. Notice that black paint stands up better than other colors. That is because it contains carbon, which is very durable. Red, on the other hand, fades rapidly. Learn to watch for this when you pass a faded sign.

Changes caused by living things

Streets that are lined with trees are very beautiful and offer welcome shade in the hot summertime. Some trees, however, have a tendency as they grow older to develop large roots near the surface of the ground. This growth is so powerful that it will crack or lift sidewalks near the tree.

Some kinds of trees are worse offenders than others. Observe the damage caused by poplars, maples, and sycamores. Pin oaks and linden trees are among the best to plant as curb trees. Examine the sidewalks near these trees.

Other damage to sidewalks is done by plants growing in cracks which have already been started. Notice how the damage keeps spreading.

Weeds can also ruin lawns and gardens unless they are promptly removed. Can you see the weeds crowding out other less hardy plants? Their fast growth causes them to take the minerals and moisture which the garden plants need.

On the other hand, when these weeds die, the minerals in them enrich the soil. You have probably seen farmers "plow under" a field of weeds.

Look around to find many wooden poles and fences which have been destroyed by rot. Of course, termites too are well known for their devastating work. One of the best ways to prevent this decay is to coat the wood with creosote before placing it in the ground. People often cover wooden posts with the black, tarry material to preserve them.

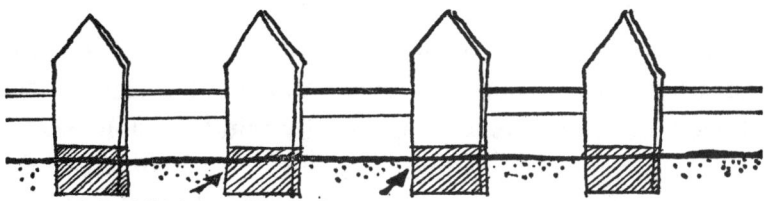

ROOFING COMPOUND PROTECTS WOOD

But there is a positive side to all the "destructive" changes that occur. Despite the cost in money, time, and annoyance for us, termites and other "destructive" living things have a definite function in this world. Can you imagine what would happen if wood never rotted? If every tree that ever grew were allowed to remain intact, what a crowded world this would be! Or, if everything were taken from the soil and nothing put back, how long could living things continue to grow?

More to find out

1. Why does an "antifreeze" solution have to be used in an automobile radiator in the winter? What kinds are there?

2. To show the force exerted by water when it freezes and expands, fill a medicine bottle to the brim, cap it, and place it in a tin can. Keep in a freezer until the bottle cracks.

3. What damage can ice do to roofs and rainpipes?

4. Find a brook with many large stones which were carried there by the water when it was moving fast.

5. Find out how fast a stream moves by throwing a stick into the middle of the current.

6. Examine painted objects which have sand blowing on them daily. Look at houses and painted signs at the beach. Are the surfaces dulled by wind erosion?

7. Find an old weathered board. See how easily the nails can be pulled out. Observe the loosened knots, cracks, and warping.

8. How are signs and awnings guarded against wind damage?

9. Why are the outside walls of a concrete foundation of a new house painted with a heavy roofing compound?

10. Put some rocks in a baking pan in a hot oven. Will expansion crack them? Carefully remove the pan to the outdoors and pour water on the hot rocks. What happens?

HOW TO LOOK AT TREES

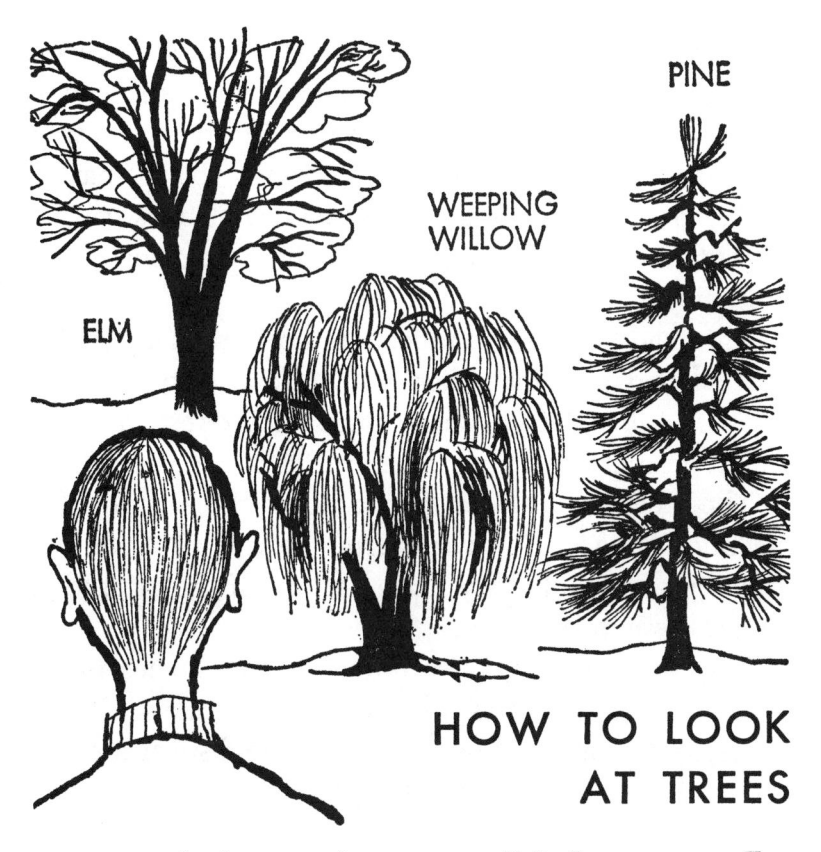

Trees are the largest plants you will find on your walks. If you know what to look for, trees can tell you many interesting stories.

One of the most practical ways of getting acquainted with the different species of trees is by seeing them over and over again. If you only learn the ten kinds of trees which are most common, you will have a great deal on which to base most of your thoughts and observations. Do not attempt to learn too many at a time, for it may be confusing.

There are many books for beginners which have remarkably clear illustrations of shapes of trees, bark, leaves, fruits, and other simple guides to identification. Most of the time a homeowner will be glad to tell you the names of the trees on his property. Ornamental trees can be identified by landscape gardeners and by catalogs from tree nurseries.

But in addition to identification there are many things to find and examine when you look at a tree. This chapter will point out many items of interest that you might ordinarily pass by during the different seasons.

The trees in your neighborhood may be useful for beauty, shade, and perhaps fruit bearing, but for our country as a whole they represent one of the most vital natural resources.

They provide the lumber for building our homes, furniture, and the countless other wooden products. Fruit, newspaper pulp, cellulose, turpentine, and sugar are just a few of a huge list of valuable products.

Trees hold the soil together, especially on slopes. Because their roots retain moisture, they reduce the dangerous run-off from rain and melting snow. They also act as wind barriers and provide homes for wildlife. Millions of people use the forest for recreation.

One of the first things you will observe when studying a tree is that the leaves are arranged to get as much sunlight on them as possible. Stand near the trunk and look up to see how this leafy umbrella is formed.

Notice that everywhere most tree trunks are in vertical positions. Even if a tree is on the side of a steep hill the trunk tends to grow straight up. In a heavily planted forest the trees grow particularly tall and straight, with the leaves only on the uppermost sunlit branches. The lower branches have died off. This is called *natural pruning* and is encouraged by lumbermen who want straight timber.

See how a big spreading tree has prevented any other tree from growing very close to it. How are the young trees under it getting along in their battle for survival?

Look around for trees which are not thriving. Try to discover the reason for their poor growth.

Some trees may have their leaves eaten away by tent caterpillars or other harmful insects. Unless sprayed quickly, these trees may die. On some of your walks you may see printed signs put up by the Park Department telling motorists not to leave their cars on certain streets because the trees will be sprayed that day.

Trees are frequently attacked by insects which bore under the bark. Gas leaking from pipes under the pavement will also kill trees.

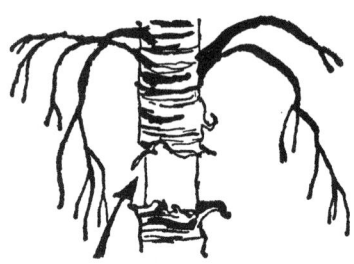

PEELING OR CUTTING COMPLETELY AROUND THE BARK KILLS THE TREE

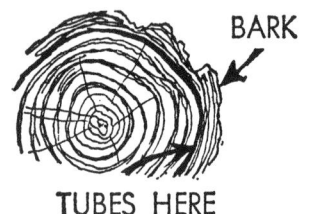

BARK

TUBES HERE FEED THE TREE

You will find many trees which have been harmed by unthinking people who peeled off bark or broke off branches. Many trees are killed by people who cut completely around a tree with a hatchet or even a knife. All the liquid food that a tree receives from the soil or sends down from the leaves goes through a very, very thin layer of tubes next to the bark. Severing these tubes starves a tree to death.

Have you ever given a

thought to how a tree gets taller? It does not grow like people. It does not increase in height from the bottom up. If two nails were hammered into the trunk of a tree exactly one foot apart vertically, they would be the same distance apart ten years later.

Trees do not stretch. Instead, only the tips of the stems grow longer. The end of a stem has a bud on it during the late autumn and winter. In the springtime the bud opens and several tiny leaves pop out. Right behind these leaves is a fast-growing shoot which makes the old stem longer. (See illustration.) After the growing season this part of the stem never grows any longer.

You can see how much a twig increased in length in one season by examining it. The distance between the two ringlike scars on the bark is the yearly growth.

IN 10 YEARS THE NAILS WILL STILL BE 1 FOOT APART

TERMINAL BUD

THE TREE GREW THIS MUCH IN A YEAR

From now on it will only increase in thickness. If you want to see where this happens, peel a little of the bark away from a live twig. The greenish layer just below the bark is the growing part of the tree. Every year it grows a little bit in diameter.

The yearly growth in thickness causes the rings which you can see on the stumps of cut-down trees. The light color of the rings is caused by the fast-growing spring growth. The dark, harder layer of wood is made in the summer.

It is fun to count the rings to find out the age of a tree. The wooden logs used for fences are good for this purpose. You can also tell which were good years for growth and which were poor years.

Do you know that the grain in lumber is caused by these annual rings? At the sawmill the tree trunk is sawed the long way and the dark summer growth gives the wood the interesting designs. If you look at the ends of boards you can still see some curved parts of the rings. A knot, of course, is a place where there was a branch in the trunk.

Another thing to do while walking is to count the age of certain evergreen trees. Look at the extreme tip of a pine tree. That lonely spire is called the *leader,* and it grows upward. From its base small side branches grow outward. Next spring there will be another leader and other side branches. By counting the number of old leaders you can come close to the real age of the tree. Some of the bottom branches may have died off, but you may still see the scars.

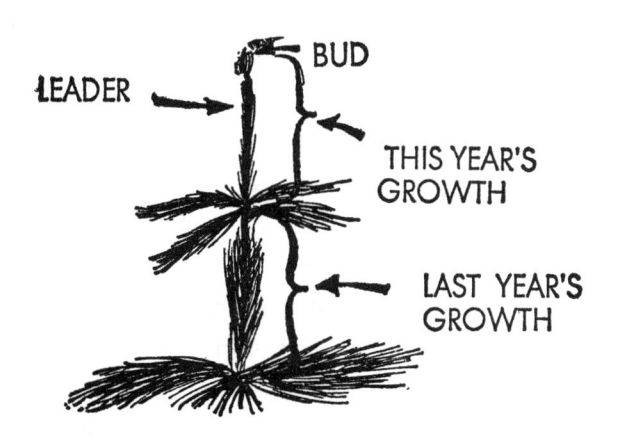

In the winter look for the buds on the leaders. Sometimes a leader is killed by the weather, disease, or birds. Then a new leader develops alongside the old one. It may grow at a slight slant. You may find a crooked trunk on an old evergreen tree because of this. Once in a while you may see a double trunk caused by two leaders developing. These are not good timber trees.

In the winter the trees may look dead, but do not think so for a moment. The buds, which were developed late in the summer, are waiting for warm weather to open into flowers, leaves, and stems. See how well each is protected. The scales are packed tightly in shingle fashion to keep out the rain and sleet. They may be glistening with a varnishlike covering. Some fuzzy insulating material may be on others.

HORSECHESTNUT BUDS ARE SHINY

MAGNOLIA BUDS ARE FUZZY

Look at them as spring approaches. They become more noticeable as they swell, and then one day—pop!—all over the neighborhood the same species will open at about the same time. However, there may be a difference of a day or so because of location. Try to account for the late bloomers. Consider the amount of sunlight and moisture, and the size of the tree. Perhaps one tree was more sheltered from the wind. Do all the trees which are under the same conditions bloom at the same time? How long do the blossoms last? Do the flowers and leaves come out of the same buds? On your walks later in the season try to remember the trees that you saw blooming. What developed from the flowers?

This is the way that trees can keep your walks interesting even though you are not a biology expert.

Leaves are green because they contain the coloring material called chlorophyll (KLO-ro-fill) that you have heard so much about. This vital substance is used by the leaf to manufacture sugar and starch from carbon dioxide and water. This manufacturing is called photosynthesis (fo-to-SIN-the-sis) and can only be done by green plants in sunlight. In the process of recombining chemical elements there is an excess of oxygen which is given off.

The water that is needed is drawn up from the roots. The gases and water vapor pass through openings on the bottom of the leaves.

Chlorophyll keeps breaking down because it is rather delicate, but it is replaced by the plant all summer as fast as it is destroyed.

But in the late summer and fall, mainly because of the weaker sunlight, chlorophyll is not produced as fast as it is being destroyed.

The leaves also contain red and yellow pigments. These cannot be seen during the growing season because the green hides them perfectly. But as the chlorophyll stops being produced, varying amounts of red and yellow become visible, depending upon the species of tree.

The final color is brown. It is the color of a dead leaf, whether it is still on the tree or on the ground.

There is a reason why a tree does not lose its sap and "bleed to death" when it loses its leaves. Before a leaf falls, a watertight, corky layer develops between its stalk and the twig. This also causes the leaf to fall easily.

Notice which trees lose their leaves earlier than others. Oaks, for example, keep their leaves for a long time—frequently all winter.

Keep your eyes on the fallen leaves as you trudge past them in the winter months. See how they disintegrate to become part of the soil.

How can anybody ever walk past a tree and act as though it were not there?

THERE ARE MANY SHAPES OF LEAVES

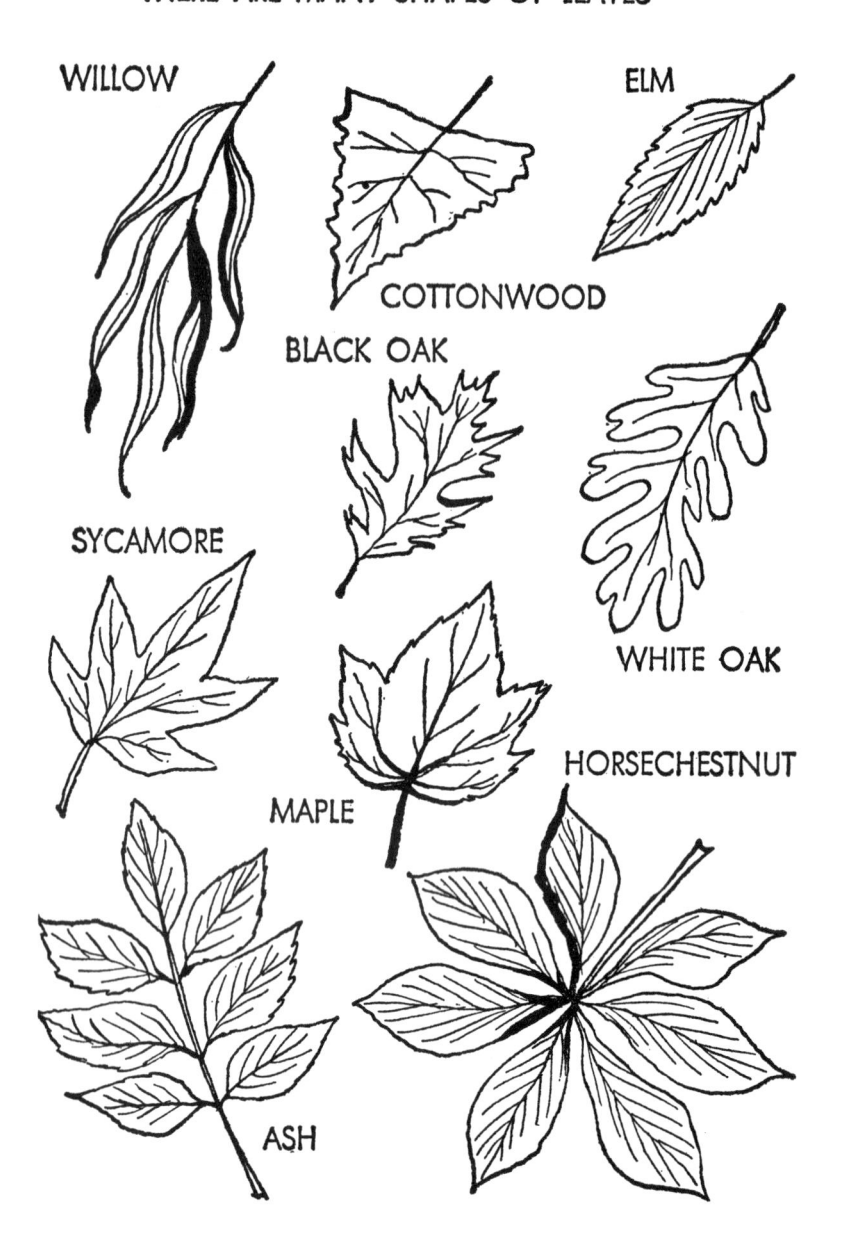

WILLOW

COTTONWOOD

ELM

BLACK OAK

SYCAMORE

WHITE OAK

MAPLE

HORSECHESTNUT

ASH

More to find out

1. Examine the work of tree surgeons in parks. Look for cement patches, reinforcement bars, etc.
2. Study the roots of a tree that was uprooted by a storm. Also examine one that was struck by lightning. What damage was done?
3. Can you find the height of a tree quickly and accurately by comparing it with the height of a person standing alongside it?
4. Learn to identify lumber by studying samples of wood.
5. Visit a sawmill, lumberyard, or wood factory.
6. Plant a maple seed.
7. Visit a tree nursery.
8. Make a collection of tree leaves, twigs, and buds.
9. Make a map showing the different trees in the community.
10. To obtain a permanent record of a tree's annual rings for home study, place a strip of paper on the stump from the center to the edge. Rub the top with a finger which is smudged with pencil lead.

PLANTS ALONG THE WAY

When you go walking along the streets in the city or small town, your attention is attracted to many kinds of colorful and interesting plants. At certain times, lawns, gardens, and fields are crowded with flowers, vegetables, vines, weeds, and other growths.

Huge volumes have already been written about this subject. This chapter, however, is a sampling of easily recognizable plants which you are most likely to find anywhere.

You can become familiar with this botanical world around you by learning just one thing at a time. By constantly meeting the same plant to which you have become alerted, you will finally get to know it as an old friend.

You will find that once you can identify a plant you will want to learn as much as you can about it.

Ask questions, read library books, and take walks with people who know about this topic. These are still the best methods for learning how to appreciate and enjoy nature's wonders.

Hedges

Hedges are shrubs which are planted to mark off walks, to hide certain areas, and for ornament. The most commonly used one is privet. Besides the green variety, you may also see a California privet with yellowish leaves. Japanese barberry is a hedge with thorns which discourage trespassing. It has red berries in the fall.

Gardens

When you pass a vegetable garden try to recognize the various plants. Many may be obvious, such as corn or cabbage. Some may already have tomatoes, peas, or peppers on them. See how they grow. Try to fix something about the plants in your mind to help you recall them next time you see them without the vegetables on them. Ask about the names of plants you do not recog-

CARROTS, BEETS, AND RADISHES

nize. When you pass a vegetable market observe some of the leaves which may still be attached to carrots, beets, and radishes. Learn about those other vegetables such as eggplant, squash, or turnips.

Many people get a feeling of satisfaction to be able to walk by a garden and know at a glance which crops are growing there and how they are developing.

Cultivated and wild flowers are with you on all your walks in the spring, summer, and autumn. Seeing them over and over will build up your knowledge of these. You may not think so, but you already know many of them. (See the lists of flowers on page 95). Do not allow an opportunity to pass when you can increase your

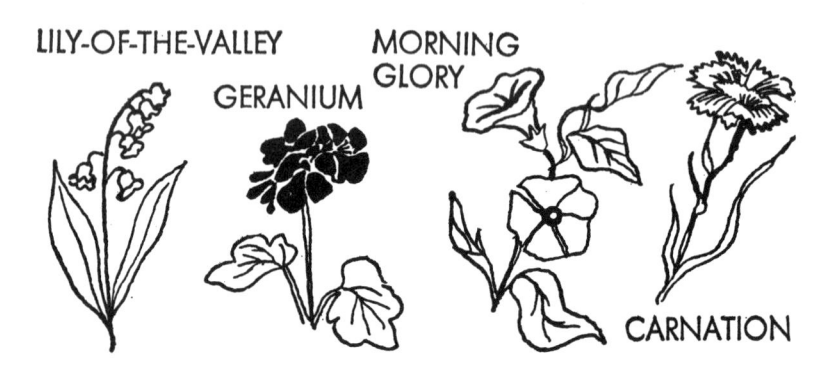

LILY-OF-THE-VALLEY   GERANIUM   MORNING GLORY   CARNATION

list. Starting with tulips, hyacinths, and daffodils in the spring, you can enjoy a colorful summer with geraniums, zinnias, marigolds, daisies, and hundreds of others. Even on your late-fall walks feast your eyes upon the asters, chrysanthemums (kri-SAN-the-mums), and goldenrod.

It is impossible in this small space to attempt to describe them. Learn those which are prevalent in your community. By doing this you are preparing yourself for an endless source of enjoyment for the rest of your life.

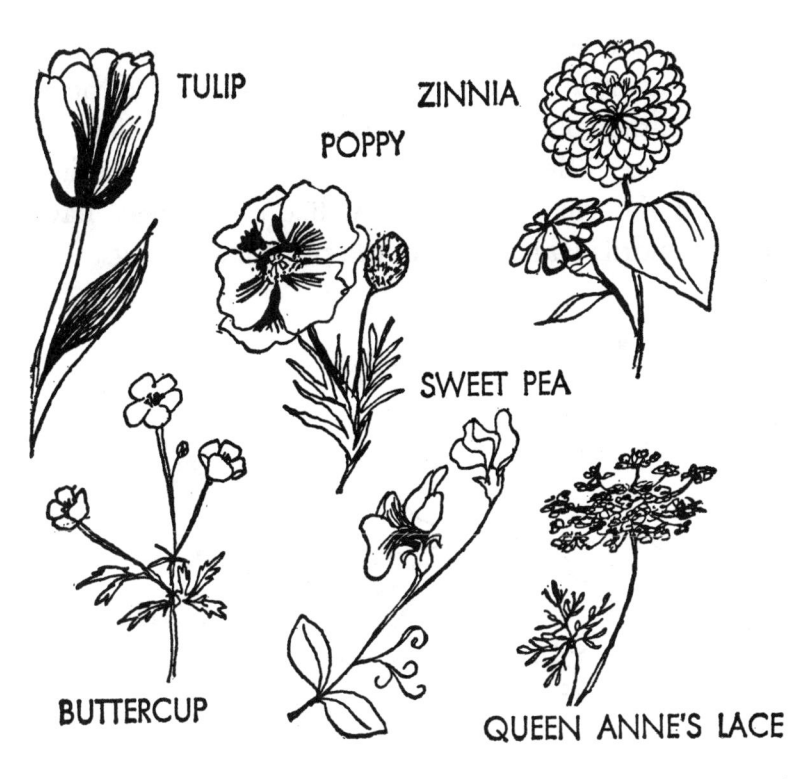

TULIP

ZINNIA

POPPY

SWEET PEA

BUTTERCUP

QUEEN ANNE'S LACE

Poison ivy

Poison ivy is a plant that can cause you great discomfort. Your best weapon is to recognize it instantly and avoid it. You are bound to find it, since it grows primarily along roadways, paths, and fences.

It is a vine with leaves which always grow in clusters of three. Each leaf is usually slightly different from the other. They may be a glossy green or a reddish brown.

Every part of the plant contains an irritating oil which can cause a painful rash, so do not touch the bark, root, leaves, flowers, or berries.

Poison ivy has little roots along the stem which cling to trees and fences. If you wish to examine other vines, nonpoisonous, to see how the weak stems can hold themselves up, then observe the ivy which clings to buildings. You may see little adhesive disks at the ends of rootlike parts on the vine. Some vines, like sweet peas, have curled tendrils which pull up the plant like springs.

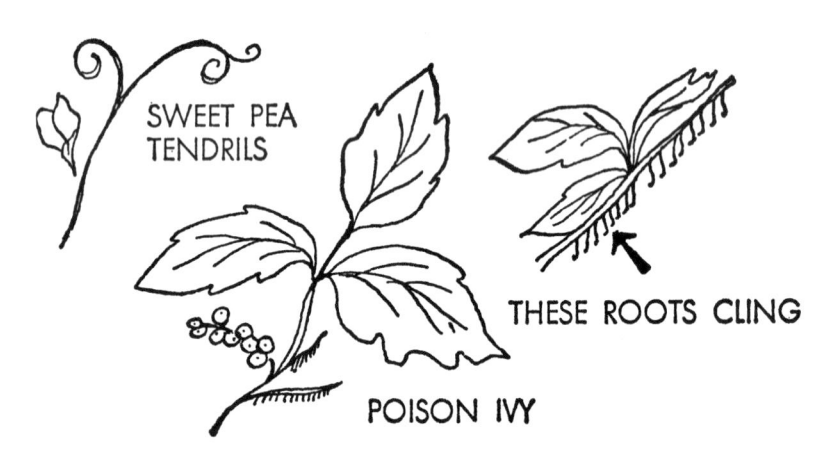

Lawn weeds

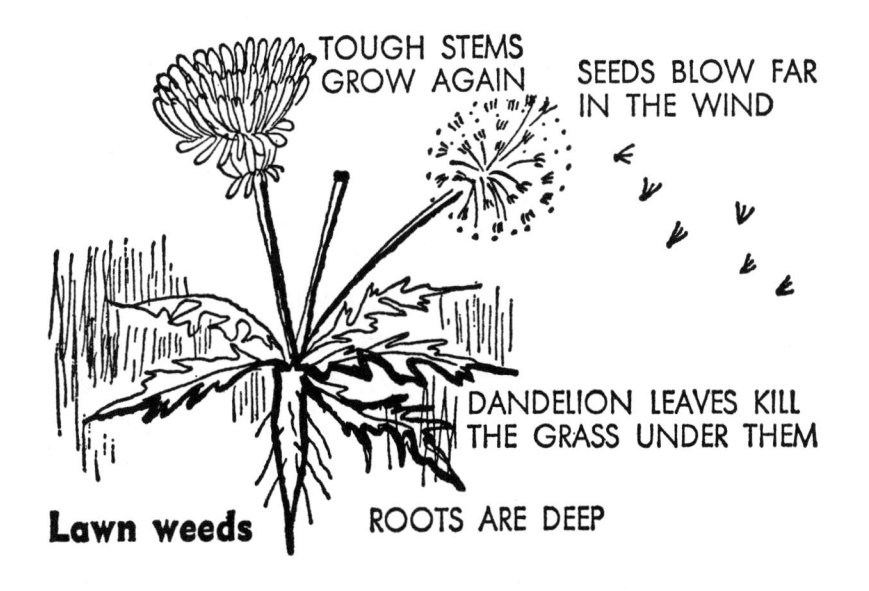

TOUGH STEMS GROW AGAIN
SEEDS BLOW FAR IN THE WIND
DANDELION LEAVES KILL THE GRASS UNDER THEM
ROOTS ARE DEEP

A weed is a hardy plant that spreads very rapidly and prevents the growth of more desirable plants. You can see the damage done by many such unwanted plants. Examine a lawn, for example, and see what is growing there besides the grass and clover.

Dandelions are bad weeds for a lawn. Their distinctive arrow-shaped leaves are flat, close to the ground, and arranged in a circle to get the most sunlight. This kills the grass underneath. In addition, the deep roots are hard to remove. When the lawn mower cuts the stems they grow again, stronger than before. Examine the yellow flowers. They are really composed of many individual flowers. You can understand now why a great many seeds are produced. See how the seeds are attached to parachutes so that they are spread over a great area.

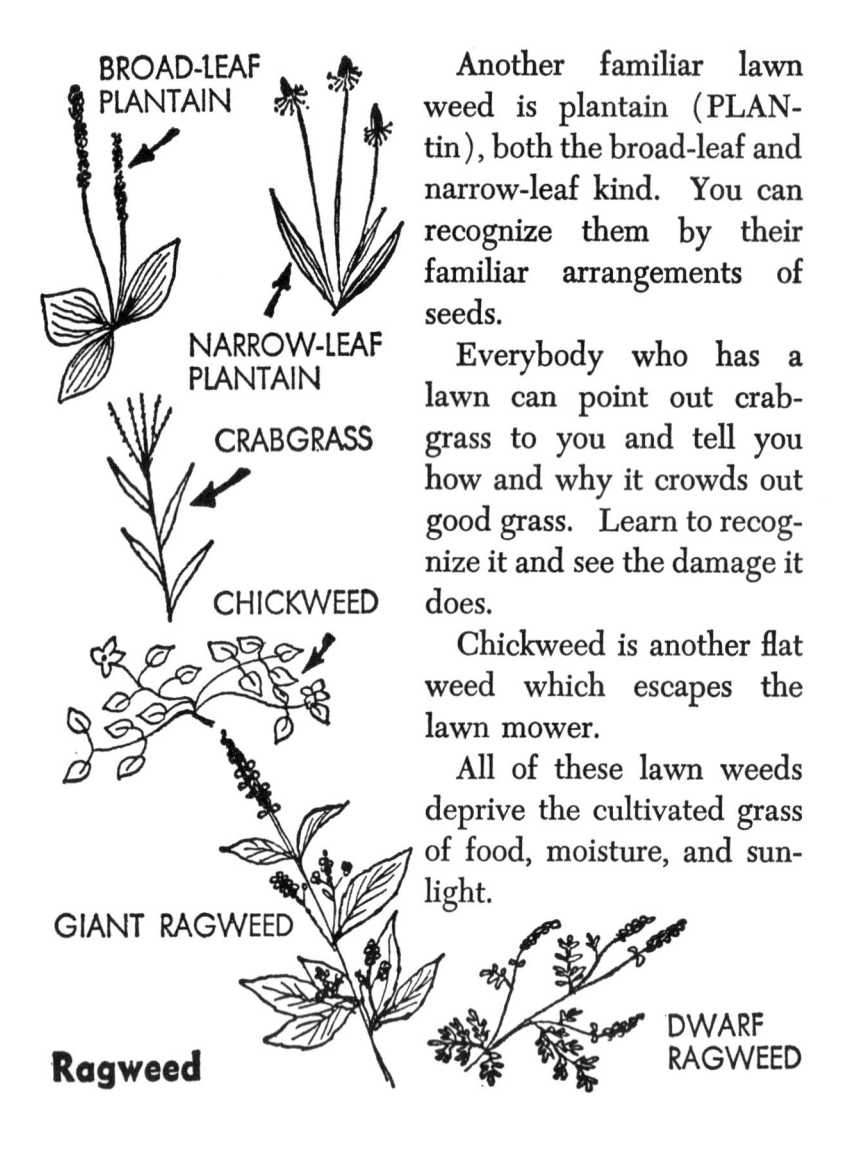

Another familiar lawn weed is plantain (PLAN-tin), both the broad-leaf and narrow-leaf kind. You can recognize them by their familiar arrangements of seeds.

Everybody who has a lawn can point out crabgrass to you and tell you how and why it crowds out good grass. Learn to recognize it and see the damage it does.

Chickweed is another flat weed which escapes the lawn mower.

All of these lawn weeds deprive the cultivated grass of food, moisture, and sunlight.

Ragweed

You may wish to recognize and know something about another common and highly publicized weed. It is ragweed, which produces pollen that is very irritating to the mucous membranes of the nose and eyes of millions of allergic people. There are two kinds of ragweed, the giant and the dwarf. These can be easily identified by their leaves as you see them growing profusely everywhere. From the middle of August until the first frost the fine pollen dust floats in the air, causing sneezes, running noses, tearing eyes, and headaches for the hayfever victims.

Goldenrod is another weed that blooms at about the same time. Many people used to blame this plant for causing hay fever. However, it hardly affects people as ragweed does.

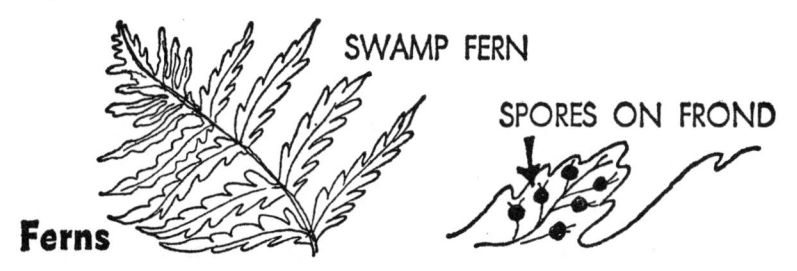

Ferns

The plants referred to thus far are seed producers, and at some time in their lives have flowers. On your walks, however, you may find ferns, mushrooms, and mosses. These are plants which never have flowers or seeds, yet they are able to reproduce more of their own kind.

Ferns are found in moist, shady places. Their divided leaves are called *fronds*. You may see how some of them uncurl from the ground in the springtime, looking like fiddle heads or thin cobras. On the bottom of some fronds you can find little dots which release tiny spores. Each spore grows into a small flat plant which produces male and female cells. From the union of these two cells new ferns grow.

Mushrooms

You probably have also seen mushrooms growing on trees, dead logs, and on the ground, and wondered about them. Those on trees sometimes look like shelves. Others on logs and on the ground resemble small umbrellas.

A mushroom is a fungus (FUNG-gus). It can live in the dark because it does not use sunlight to make its own food. In fact, it looks so pale because there is no chlorophyll in it. It must get its food from other plants, and for this reason is called a parasite. Most of the mushroom plant is hidden under the ground or in the tree.

These fungi (FUN-jeye) are useful because they break up dead matter into minerals which become part of the soil.

Many mushrooms are good to eat, but since some are poisonous, never take any home to be cooked. To be safe, buy them in the store.

The mold on rotting logs or under stones is also a fungus. All fungi reproduce by means of spores. So does moss, that spongy, carpetlike fungus which grows in moist shady places.

Lichens

Sometimes you will discover scaly, dry-looking, gray or green patches on trees or rocks. These are caused by lichens (LIE-kins), which consist of two plants, a fungus and an alga. An alga contains chlorophyll and can make its own food, which the fungus cannot do. However, the fungus can absorb and hold water, supply minerals, and also anchor itself. Each contributes its useful part to this interesting partnership so that both can survive.

More observations

There is always something to think about or to do on a nature walk. See how the mound of earth dumped by a truck two weeks ago is already being covered by green plants. How did they get there?

Observe how poorly the lawn grass grows under some shade trees. Discouraged gardeners frequently will grow ivy or other shade-loving plants in such places.

Notice how particular most plants are concerning their special growing conditions. Some only grow in wet places, some in dry, some like sunny spots, others shady. Once you know a plant well, you can almost tell where you are going to find it.

Which plant attracts most insects?

Does the sunflower turn so it faces the sun all the time?

Did the blackberries and huckleberries ripen since your last walk?

PORTRAIT ATTACHMENT

On your walks in the summer or autumn, can you recall what the same path looked like in the spring?

Do you frequently crush the leaves of mint and sassafras, or the bayberries, between your fingers to get their enjoyable aroma?

Which plants have seeds that stick to your clothes?

Have you learned to use your camera to capture the beauty of flowers on film?

Do you think of how natural wild spots in your neighborhood are disappearing as new houses are being built? Should the town preserve a few places here and there as wildlife refuges, before they all disappear?

These are some thoughts to consider as you walk.

A word about picking wild flowers. If you wish to gather some flowers for identification or preservation only pick those which grow in great profusion. If it is just a single flower you should leave it. In fact, some states protect certain wild flowers. Find out about that.

Enjoy but do not destroy!

A beginner's list of familiar flowers

Garden flowers

Alyssum	Marigold
Aster	Morning glory
Begonia	Nasturtium
Chrysanthemum	Pansy
Cosmos	Poppy
Daffodil	Petunia
Daisy	Phlox
Day lily	Rose
Four o'clock	Salvia
Geranium	Snapdragon
Hollyhock	Sunflower
Hyacinth	Sweet pea
Iris	Tulip
Lily of the valley	Zinnia

Wild flowers

Aster	Jack-in-the-pulpit
Black-eyed Susan	Jewelweed
Butter and eggs	Milkweed
Buttercup	Morning glory
Chickory	Mullein
Clover	Queen Anne's Lace
Daisy	Self-heal
Dandelion	Skunk cabbage
Day lily	Thistle
Goldenrod	Violet
Hawkweed	Water lily
Honeysuckle	Yarrow

More to find out

1. How many different kinds of vines can you observe which grow on buildings?
2. Make a collection of different seeds that you find on trips.
3. What causes the dew on grass?
4. Find out which common vegetable juices were used by the Indians for dyes.
5. Grow mold by placing a piece of bread in a moist, dark, and warm spot for several weeks. Examine with a small magnifying glass.
6. Make a collection of many kinds of ferns. Press and dry each fern between flat sheets of newspaper.
7. Try to go on field trips with groups from high schools, clubs, and museums.
8. Find the difference between wheat, oats, barley, rye, buckwheat, and timothy.
9. Have a treasure hunt for wild grapes, last year's goldenrod stalks, cattails, blackberries, vines with five-leaf clusters, orange-colored mushrooms, wild strawberries.
10. Ask your doctor for the most up-to-date remedy for poison ivy.
11. What is the latest treatment for hay fever?

ANIMAL NEIGHBORS

If you walk where there are houses you will only see such familiar animals as cats, dogs, horses, and perhaps cows or goats. These are called domesticated animals because they have become adjusted to living under conditions established by people.

As you pass fields and woods you may be fortunate enough to see some rabbits, squirrels, chipmunks, snakes, and others. These wild animals are rather shy and know how to hide. Some are most active in the evening, so you may rarely see them. But every animal reveals its presence by telltale signs such as gnawed food, tracks, droppings, and location of its special kind of shelter.

It is fun to read the story of an animal's activity just by learning to observe more carefully what you may have been overlooking before. A few ideas are included here. There are many books in the library entirely devoted to this.

When many people speak of animals they only think of the furry kind. From a scientific viewpoint, however, an animal is any living thing which is not a plant. Therefore earthworms, insects, fish, birds, snakes, and frogs are animals. Man himself is listed in the animal kingdom.

Animals with hair or fur are called mammals. These are warm-blooded and nurse their young by means of milk glands.

No matter what they look like or where they live, all animals are busy most of the time finding and consuming food and water. They also must have some kind of shelter in which to hide and rest.

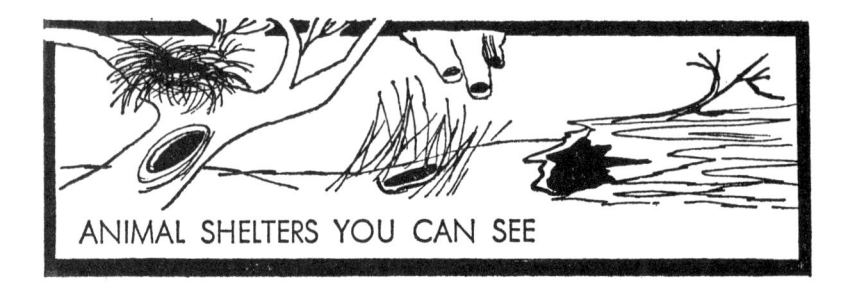

ANIMAL SHELTERS YOU CAN SEE

Dogs

Dogs are probably the most common mammals in residential sections. They have been around people since the ancient cave men used them for hunting and for food. Some zoologists believe that they originated from wolves.

There are many interesting observations to make as you walk past a dog. Can you name the breed? If not, ask the owner. If the dog walks beside you, listen to the clicks of his claws on the sidewalk, especially if it is a large dog. A dog does not pull back its claws the way a cat does. If you can study clear dog tracks on soft, smooth earth, observe the claw marks that are typical of a dog.

CAT TRACK

DOG TRACK

Notice how a dog sniffs at everything with its very sensitive nose. See how his ears are constantly twitching and standing up. A dog can hear distant sounds better than we can, often high-pitched ones which are beyond our range.

Dogs have weak eyes and depend upon other highly developed senses. They have wider angular vision and can see about 250 degrees around them when looking straight ahead. We can only see about 180 degrees this way. Dogs are color-blind and only see things in white, black, and shades of gray.

Watch a dog pant with his mouth open on a hot day. Drops of water may drip from the tongue. A dog's skin does not contain many sweat glands like ours. Therefore a dog cools off by perspiring through the tongue.

When you see a dog chasing a cat, do not believe that it is a dog's instinct to do this. Psychologists have discovered that a dog learns to do this by watching other dogs. Of course, how the first dog learned is still a mystery.

Cats

The familiar house cat that you see everywhere in town is a relative of the jungle lion. In fact, the Latin name for a cat is *Felis domestica,* while the lion is *Felis leo.*

Observe how smoothly and slowly it walks near shrubs, always ready to stalk a bird it may spot ahead of it. Can you see or hear whether the cat has a little bell on its collar to prevent a stealthy approach?

Sometimes a cat will play with a dead bird or mouse before eating it. You may even witness a mother cat teaching a kitten how to kill mice. Kittens are not born with this ability. Some kittens are even afraid of mice.

Stop and watch how a cat washes its face and body by licking its paw and using it like a washcloth. Cats are very clean house pets.

Make a very soft whistling sound and watch how the ears of a cat show that it has caught the sound.

A cat can see in very dim light, but of course it cannot see in absolute darkness. Did you ever see how the eyes of a cat light up at night when a car's headlights strike them? This is caused by a reflecting layer of cells in the eye, which can concentrate the dimmest glimmer of light.

A cat can also open its pupils very wide in the dark to allow as much light as possible to enter. On the other hand, in bright sunlight the pupils close down to narrow slits.

Sometimes you may see a cat use a tree for a scratching post upon which to sharpen its claws. Look for these scars on young trees. A cat is a carnivorous (meat-eating) animal and likes to keep its claws sharpened.

Cats have well-developed instincts, and they may wander far from home without losing their way. That is why you may find a house cat out in the woods quite a distance from its home.

Squirrels

Squirrels are wild animals that you can observe first-hand in the city. You will usually find these rodents where there are trees which produce acorns, nuts, or cones. They are very busy gathering food to bury or hide for the winter or to eat at leisure. See how they carry the nuts in their roomy cheek pouches.

Sometimes you may find a cluster of broken shells on a flat tree stump. Squirrels like to use one particular spot as a dining table. Pick up some acorns to see the teeth marks. Squirrels also eat berries, flower buds of trees, and other soft foods. Many of the oak trees were actually planted by these animals, who forgot where they buried their acorns.

Note how many ways a gray squirrel uses its tail: for balance, for a steering rudder when making long leaps, as a blanket, and for showing his feelings. A red squirrel's tail is not as bushy.

Look up among the highest branches and you may see a squirrel's nest made of twigs and leaves. It is as big as a bushel basket. These little animals may also live in holes in trees.

Squirrel calls may sound like those of birds until you become familiar with these loud scolding notes. They sound like "yak-yak" repeated many times.

Chipmunks

A chipmunk is easily distinguished from a squirrel by the stripes running along its back. It lives in burrows, where it stores many nuts and other food. Because of this, a chipmunk comes out of hibernation in the spring noticeably fatter than other animals.

Like squirrels they are rather friendly, and one may walk along with you for a while. Their call is a soft "chuck-chuck."

Rabbits

Cottontail rabbits, with their white powder-puff tails, can be seen on almost any walk you take close to town. Many will come right into your back yard and eat some of your vegetables or the bark of very young trees. The wire screening you often see around little trees is to prevent such damage.

When frightened, a rabbit will "freeze" and not run until the very last moment. The special eyes on each side of its head can see in almost a complete circle without making the slightest head movement. Notice how the ears, which are very flexible, turn to face the sound.

A female rabbit is called a doe, the male is called a buck. The young matures rapidly; in two weeks after birth it is ready to go on its first trip alone.

Rabbits are usually quiet, but when attacked they can give out heart-breaking screams. A rabbit's way of signaling trouble is to thump its hind feet on the ground.

In the snow you can measure the length of a rabbit's leap. Normally it is about four feet long but when there is danger it may be more than ten feet. Rabbits are very fast runners for about 80 yards, after which they tire.

Horses

The horse that you can still see in some sections has had its ancestry traced back 60 million years to a small animal about the size of a cat.

Did you know that up to the age of five years, a male horse is called a colt and a female is a filly? After that they are known as stallions and mares.

A horse sleeps when he stands. This is more comfortable than lying down, because his heavy weight makes it difficult to breathe; it also cramps the muscles. When standing still, the leg joints automatically lock.

Skunks

There are few people who would not recognize the glossy black body and white stripes of a skunk, nor is there anyone who has not smelled its powerful odor. When frightened, skunks release a strong, unpleasant smell from their scent glands. Often, when walking, we can smell a skunk who has been killed while crossing a road. This is because they move only slowly and cannot get out of the way of a fast-moving automobile. Skunks can be "deodorized," however, by removing their scent glands, and make good pets.

Toads

A toad may hop across your path. Do not confuse it with a frog. Frogs have moist, slimy skins, while toads are dry and warty. Toads are slower moving, more clumsy, and heavier in appearance, and usually are on dry land. Frogs are rarely far from water.

A toad has a bump behind each eye containing a gland with a bad-tasting fluid which its enemies dislike. It has never been proved that people get warts by handling toads.

Frogs and toads are useful because they are skillful at catching many insects with the long sticky tongues that are attached by the tip to the front of the jaw.

Salamanders

The little red salamander you often see in moist, shady woods is called a red eft. It will become a newt when it returns to the water in one or two years. Salamanders and newts should not be called lizards, since lizards have scales. Salamanders do not.

Despite the fact that the building of houses in the suburbs is removing the natural habitats of wild animals, there is still much to be seen if one knows what to look for.

105

More to find out

1. Learn to recognize tracks made in soft earth and in snow by cats, dogs, field mice, rabbits, squirrels, small and large birds. Try to piece together a story just from the tracks.

2. Find out from veterinarians how to get the age of a dog.

3. Why are earthworms considered good for agriculture?

4. Find out how a bat can fly around obstacles in complete darkness by means of its built-in radar system.

5. How can you tell that there are deer in your neighborhood?

6. Learn to recognize a few common snakes that you find on your walks. Visit a zoo.

7. Can you identify the breed of every dog that you see?

8. Do you know the call of a bullfrog?

9. Do you know the difference between a box turtle and a painted turtle?

10. Keep a list of all the animals that you have seen on your walks. Can you give an interesting five-minute talk about each one? Read some of the fascinating books that are suggested.

BOX TURTLE   PAINTED TURTLE

HOUSEFLY — HEAD, THORAX, ABDOMEN

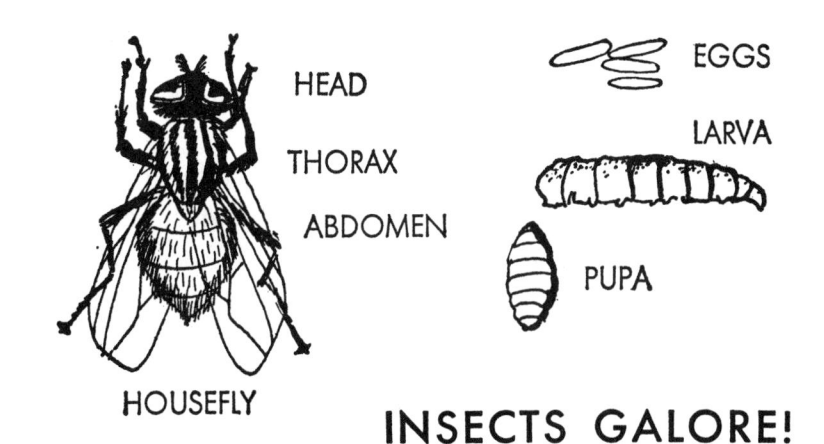

EGGS, LARVA, PUPA

INSECTS GALORE!

There are more insects in the world than all the other animals put together. Experts say that there are surely several million *kinds* of insects. It is no wonder that you see them, feel them, hear them, or notice their work on every walk you take. They are in the air, water, soil, wood, trees—in fact, almost everywhere!

An insect has three body sections: a head, thorax (chest), and abdomen. It has three pairs of legs and usually two pairs of wings. It has no bones, and the outside hard crust is its skeleton. It breathes through tiny openings in its abdomen and thorax.

Sometimes the young insect does not emerge from the egg looking like its parents. Instead, it comes out of the egg as a wormlike larva. It usually will eat only a very particular food, grow fat, split its sides, and become a larger larva. After a time it stops eating and forms a covering around itself called the pupa (PYOO-pa) case. Inside, it develops into an adult which will emerge from the pupa case. This remarkable change is called *metamorphosis* (meta-MORE-fo-sis). Each insect has its own peculiar way of doing this.

The topic of insects is so huge that only a few observations will be suggested for several of the most common insects. It is hoped that more interest will lead you to some good library books.

Moths and butterflies

Can you tell the difference between a moth and a butterfly? There are many differences, of course, but the best way is to look at the feelers, or *antennae* (an-TEN-ee). If they are like threads with knobs or bulges at the ends, the insect is a butterfly. Moths have feathery antennae with no knoblike swelling at the ends. Moths usually have thicker bodies and fly at night, while butterflies are seen mainly in the daytime. Moths spread their wings out flat when resting, while butterflies usually fold them vertically over their backs.

A BUTTERFLY HAS KNOBS ON ITS ANTENNA

A MOTH HAS FEATHERY ANTENNA

A BUTTERFLY CATERPILLAR MAKES A CHRYSALIS

A MOTH CATERPILLAR SPINS A COCOON

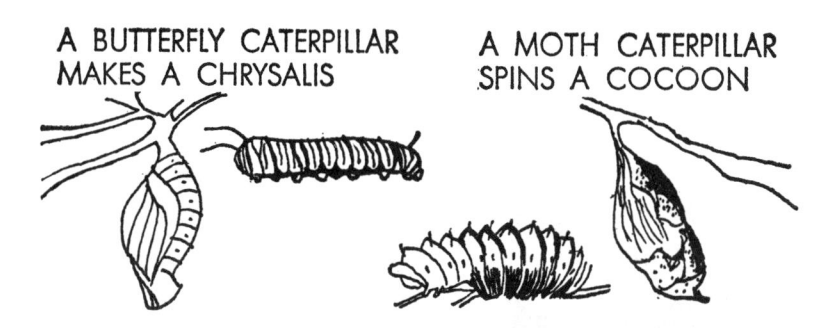

The larvae of both are the familiar caterpillars. However, a moth caterpillar spins a cocoon in which it pupates and becomes an adult, while the caterpillar of a butterfly forms a tight covering about itself called a *chrysalis* (KRI-sa-lis).

When the adult emerges from its cocoon or chrysalis, it stretches out its folded wings. After a while the wrinkles disappear and the veins harden permanently. This gives structural support to the delicate wings.

Once the butterfly or moth emerges, it does not ever grow any larger. The small and large butterflies you see are different species.

The powder which may come off the wings onto your hands is really made up of scales which overlap like shingles on a roof. Observe them with a magnifying glass.

When you see a moth or butterfly at a flower, it may be feeding. It does this by unfurling a coiled-up tube through which it sips nectar. Catch a butterfly and look for this sucking tube. Some of these insects are born with no mouth parts for eating, and die of starvation soon after mating.

If you find a moth's cocoon, see how it is attached to the branch. At home, cut it up carefully to study its waterproof and insulated construction. Watch the pupa move when placed on your warm hand.

Mosquitoes

The pesky mosquitoes which swarm around you on your walks and bite you are all females. The males live on plant juices and never suck blood. The humming noise you hear around your ears is caused by tiny parts vibrating in the mosquitoes' air passages.

Mosquitoes abound in areas which have marshes or other quiet water, because the eggs are laid in water in neat rafts. On hatching, the larvae hang suspended from the surface of the water, breathing through tiny tubes. Then they curl up and turn to pupae, from which the adults emerge.

Spreading a thin film of oil over their breeding places kills the mosquitoes by closing their breathing tubes. They may also be controlled by draining swamps, using DDT, and stocking water with fish that feed on mosquito larvae.

If a mosquito bites you, try to remember that those which carry malaria and yellow fever thrive only in very warm climates.

Flies

Look at the house flies feeding on garbage and other rotting materials. Soon they may be on your food, unless you prevent them from entering your home. Flies lay eggs in exposed refuse. The larvae are called maggots and are white, moist, and resemble worms.

House flies cannot bite you because they have only sucking mouth parts. In order for a fly to eat something solid, like sugar, it has to wet it first by means of the undigested juice from its last meal in its own stomach. Only then can it suck up the solution. This is what spreads disease. It is also the reason for the fly specks in the kitchen.

Flies have very hairy legs which also carry germs. There are suction pads at the ends of the fly's legs which enable it to walk on ceilings and walls.

What happens to flies in the winter? Most of them die, of course, but enough survive to start new generations next spring. They find shelter in cracks, under bark, or in attics of homes. You must have seen a fly in your home in the dead of winter.

Dragonflies

The beautiful dragonflies darting past you eat mosquito larvae which float on the water looking like rafts. They also catch flies and mosquitoes which may be flying. Their long pointed bodies have earned them the name "darning needles," but they can never harm you because they have no stingers.

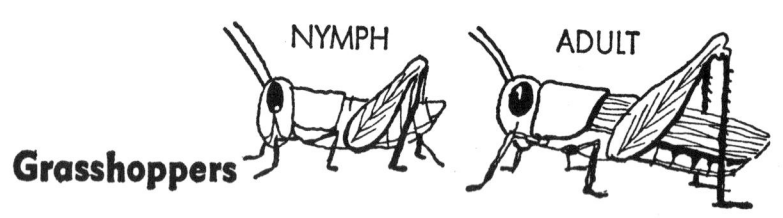

Grasshoppers

You will see many different sizes of grasshoppers. One of the reasons for this is that grasshoppers do not go through the kind of development that moths do, where each stage looks so different from the other. Instead, grasshoppers' eggs produce young which, from the start, look like their parents, except that they are smaller and without wings. They are called *nymphs*. After growing and molting (splitting their skins) several times, they develop wings and become the size of adults.

A grasshopper makes noise by rubbing the rough part of its rear leg against the cover of its wings. But, strangely enough, the ears are located on both sides of the forward part of its abdomen!

Ants

Many people like to watch ants bringing materials to an anthill. Stop and study all the activities and see the display of strength and intelligence. Another way to find ants is to turn over a large flat stone or board. You will see how the ants start rushing around, each one carrying an egg or a cocoon which looks like a grain of rice. If you poke under the surface with a stick you may see ants in all stages of development.

Usually, only the queen and the males have wings. The queen ant is larger than the others and is busy laying eggs all the time. There are very few males; almost all the ants you see are females. They do all the labor. Males never work.

Ants are called social insects because they live in communities. The work each one does seems to benefit the entire colony.

Termites

Break apart a rotting log and you may see termites, with their pale bodies, working in the dark. Termites chew into the wood and swallow it, but oddly enough, they cannot digest it. They would starve to death were it not for certain microscopic animals called *protozoa* (proh-to-ZO-a) which live in the digestive systems of the termites. These protozoa are able to change the wood into sugar, thus supplying nourishment for themselves as well as for their hosts.

Fireflies

A firefly is quite easy to catch and examine. It is called a glowworm, but the hard wing covers which also go over the top of the abdomen classify this insect as a beetle.

It is the male which does the flying. The female has no wings and is in the grass or on the ground. She has a much brighter light than the male and probably tries to attract him down to her by flashing at a certain rate.

The cold light comes from the tip of the abdomen. Fill a jar with fireflies, hold it near a newspaper in the dark, and see if you can read by firefly light.

Praying mantis

A praying mantis is like a dinosaur, compared to most insects. It gets its name from the fact that it rests on its four rear limbs and bends its two front legs as though in prayer. There are spines on these front legs for holding and cutting up the insects which it catches. A praying mantis looks like a twig when it stands still.

It is the only insect that can turn its head and look over its shoulder the way man can. Most insects have the head and chest fused together. Use a magnifying glass to study its head and its interesting jaws. After mating, the female squeezes a white froth from her abdomen. She lays her eggs in this egg mass. You may find the hardened egg mass of this useful insect attached to a twig. It is about an inch in diameter and light brown in color. Keep it in a jar with a pierced cover, and in the spring it will hatch into a few hundred young insects looking like the parent but without wings.

Water striders

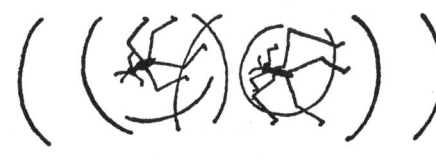

Water striders are those long-legged insects which dart around on the surface of still water looking for food. They can do this for the same reason a needle will float upon water. Water seems to have a weak film over it which we call *surface tension*. The water strider's light weight and its six broad feet, which distribute its weight evenly, prevent it from breaking this film.

Bees

BEE STING

The bee that you see so busily crawling into flowers collecting nectar to make honey is a female. She actually works herself to death in a few weeks after she is born. See how the pollen sticks to her body. Some of this will rub off when she goes to another flower. In this way she helps in the production of new plants.

You should have respect for her sting, which she uses only when frightened. The sting has barbs on it, so that it is usually impossible for the bee to remove it from your skin.

In her frenzy to escape she tears out the soft portion of her body which is attached to the sting, and she dies.

Do not attempt to remove an imbedded sting yourself. You may leave pieces behind and develop an infection. A doctor has a special technique to remove a bee sting safely.

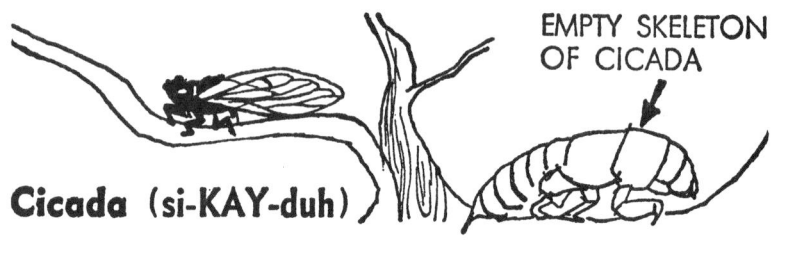

Cicada (si-KAY-duh)

EMPTY SKELETON OF CICADA

The sudden loud rasping sound you hear on your walks is probably produced by a cicada. There are two tightly stretched membranes on each side of its abdomen, which are attached to muscles. The male cicada makes these drums vibrate to produce the noise.

On your walks you may sometimes see the empty skeleton of a cicada attached to the trunk of a tree. This was split open when the cicada outgrew its old skeleton. The process is called *molting* (MOHL-ting). A bigger and perfectly formed cicada emerged and flew away.

Spiders

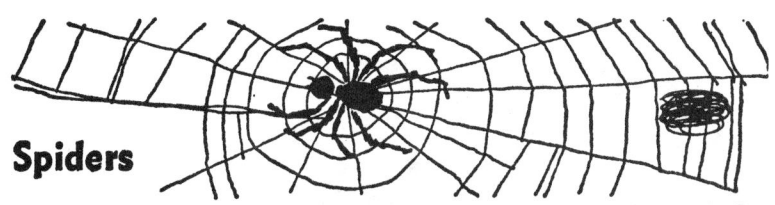

Sometimes you walk right into a cobweb and spoil the work of a spider who was waiting to pounce upon an insect that got tangled in the sticky web. Spiders have eight legs and are not insects, but are their distant relatives, called *arachnids* (a-RAK-nids). An interesting thing to do next time you pass a fresh spider web is to throw an insect into it. In a twinkling the hidden spider will be upon it.

Centipedes and millipedes

Centipedes and millipedes are not insects, although they are related to them. They do not have as many legs as the names "hundred-leggers" and "thousand-leggers" suggest.

It is easy to tell the difference between these two animals, because the centipedes have one pair of legs for each segment. Millipedes, however, have two pairs of legs for each segment. Centipedes are useful because they eat household insect pests like roaches, bedbugs, and flies. Most centipedes are not poisonous, although the bites of large centipedes in the South and in the tropics can be painful.

Millipedes are harmless. They eat rotting wood and decaying vegetation.

More to find out

1. What damage is caused by tent caterpillars? Observe how these blue-gray caterpillars with white lines on their backs make their tents. They spend the night in them. During the day they are out feeding.
2. Look for green caterpillars in cabbage leaves. They are larvae of the white cabbage butterfly.
3. Find and study a large green tomato "worm." It is the caterpillar of a sphinx moth.
4. Collect cocoons in the winter. Keep indoors in jars with pierced covers. Sprinkle some water in once in a while. See what emerges in the springtime.
5. Visit a beekeeper in the neighborhood.
6. House flies cannot bite, but female green-headed horse flies do. Find out which other flies bite.
7. How many different kinds of insects can you find in your back yard?
8. Learn how to catch, kill, and mount butterflies.
9. How are harmful insects controlled by farmers?

GETTING ACQUAINTED WITH BIRDS

It is always exciting to watch birds even if you are not in the bird watchers' fraternity. Walking along, you will discover that there are more birds around than you thought. It can be a source of great interest to study their behavior. See if you can learn to name the most common birds in your community. Before you are aware of it you will also become familiar with their songs and calls.

If you become friendly with a bird watcher and go for several walks with him, you will probably learn more than you can by yourself. In a short time you will know what to look for when you go walking alone. You can also obtain excellent inexpensive bird guides to which you can refer.

Birds are busy getting their food from early morning until dark, and require a great deal of energy to fly and maintain their temperature of about 106° Fahrenheit.

(Ours is 98.6° F.). This keyed-up existence is noticed in everything birds do. Observe the fast motions as they feed and how quickly they fly away at the slightest movement.

You will find that birds have very strong preferences about certain foods. In fact, the bills, legs, claws, wings, and other parts of their bodies are specially built for obtaining seeds, fish, mice, insects, and other foods. That is why you will usually see certain birds in marshes, woods, meadows, garbage dumps, seashore, or swooping tirelessly through the air with their mouths open to catch insects.

Watch how a bird drinks water. It stands at the edge of the water and dips in its bill. Then it puts its head back so that the water trickles down its throat.

On a warm day you can see a bird take a bath. First it goes into the very shallow water. Then, by means of great activity with its wings, it splashes water around its sides. It probably does this to cool off or to loosen some irritating insects.

Ahead of you on a dirt road you may see a bird taking a dust bath with the same motions as for a water bath. This may be freeing it from lice, mites, and other insect pests.

Watch a bird landing on a branch. See how its wings and tail open up to act as air brakes, just like the flaps on an airplane. Its tail immediately flips to balance it, like the parasol of a tightrope walker. All this occurs very quickly, and most people miss this interesting technique.

Once a bird is perched, only an occasional flick of the tail is necessary to balance it. The bird can even go to sleep without falling over because its weight automatically tightens the claws around the branch.

When you see a bird preening itself, watch how it uses its bill. There are oil glands at the base of the feathers. Spreading the oil over the feathers makes them waterproof and glossy. At the same time this action hooks together ruffled feathers, which engage just like a zipper.

It is difficult to come close to a bird because it trusts no one. It will fly away from any movement or sound, even a harmless one. Better to be safe than sorry. Birds have sharp eyes, which can spot objects from a wide angle. Their ears cannot be seen but are holes on each side of the head, covered by whorls of feathers.

In the city the worst enemy of birds is the cat. Cats should wear collars with tiny bells on them so that the birds can be alerted. Food should never be placed on walks or feeding platforms where a cat can stalk and pounce upon a bird.

When a bird flies, its wing shape changes too quickly

for you to see how it does it. Scientists using high-speed cameras have demonstrated that the wings change curvature and outline on the up-stroke and down-stroke so that they act like propellers, causing forward motion and also providing support.

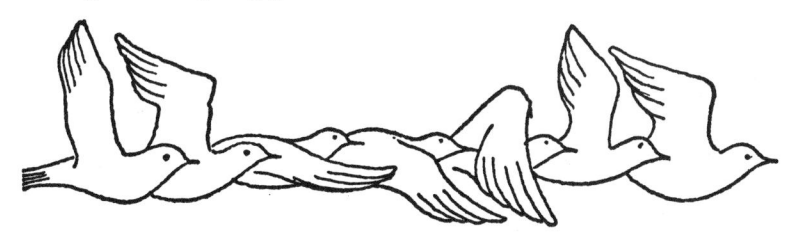

Watch a large bird soar without seeming to move a muscle. It can do this because of air currents against it. You may be able to see it use the tail for steering.

The streamlined shapes of birds help them to fly. They also have hollow bones and light feathers. You never see birds panting as other animals do when they are tired, because birds have large lungs and also carry reserve air in sacs outside the lungs and in the hollow bones.

Birds can exchange the food they eat for muscular energy very quickly, because they have a highly efficient, short digestive system. That is why birds eliminate so frequently.

Most of their melodious singing takes place in the spring and early summer, but there are many other songs and calls that you will hear on your walks. The reason for their singing is still a mystery, but many bird experts believe that the bird is proclaiming to other birds that the territory is staked out for him. Perhaps it is a warning that there is only enough food for one bird and his family.

You can often hear alarm cries of birds. Sometimes when an enemy is in the vicinity a robin will start scolding, and soon many robins in the vicinity set up a noisy squawking. Sometimes the birds dive bomb the intruder, who gets discouraged and leaves.

On your walks you will hear crows answering each other. You may be witness to a noisy display of gang warfare when many crows team up to chase away a hawk, an owl, or even a fox.

Sometimes you will see two differently colored birds together. A good look will show you that they belong to the same species and are male and female. Among the birds, in almost every case the male is the more colorful one, and it is he that does most of the singing. The female ordinarily chirps.

If you see an agitated bird with a worm or insect in its bill, you can be almost sure that there is a nest with young nearby. The bird will not go into the nest when there is any danger. In fact, it may draw you away from the nest's location. It may even behave as though it has a broken wing or is hurt, in order to lure you away from its young. But if you remain motionless from a reasonable distance, you can trace the bird to its nest and see how the parent feeds the infants. Learn to recognize nests made by different birds. Look for unusual places for nests in the city—in street lights, under awnings and eaves.

In addition to their general characteristics, there are individual things to observe in specific birds.

SPARROWS HOP

STARLINGS WALK

DOES A ROBIN WALK OR HOP?

English sparrows are hardly ever found very far from houses. They are real city folks.

Notice that a pigeon and a starling walk, while a sparrow always hops. Does a robin walk, hop, run, or combine all three?

Watch a robin move on a lawn, suddenly stop and stand at attention like a soldier. This action enables you to pick out a robin at a distance.

Notice how a robin appears on a lawn which is being watered. The bird is looking for the worms which come to the surface. See how it sometimes braces its legs and tugs to pull a worm from the ground.

When you hear the thumping of a woodpecker's chisel-like beak against the trunk of a tree, approach and watch how it works. Watch how it uses its sharp, stiff tail feathers for propping up its body. Observe how the four toes on each foot (two above and two below), help it cling to the bark. Later, look at the chips at the base of the tree. See the holes the bird made in the tree looking for insects to spear with its long barbed tongue. No one can explain how the bird seems to know that there are insects under the bark. Woodpeckers like to work on diseased or dead trunks, for they seem to know that these trees have many hidden insects.

You may see some pheasants in the fields on the edge of the woods. You will probably find that only one has the long colorful tailfeathers. This is the male, who generally has a half dozen females in his "harem."

The birds which keep flying for hours in graceful arcs without landing are probably swallows. They are insect eaters and fly with their mouths open, scooping up hundreds of flying insects for their meals. When they are not flying they rest with many of their kind on electric wires along the road.

If you see a large bird gliding in lazy circles or hovering in one spot overhead while its wings are moving rapidly, it is probably a hawk. Its telescopic eyes can pick out mice and small birds from quite a distance overhead. In a screaming power-dive, it falls on a victim with its sharp talons.

If you live near the ocean you can see terns hover over the water and suddenly dive, coming up with a small fish or other tidbit to eat.

Gulls have weak bills and so are adapted to eat soft things such as garbage, dead fish, and other sea life which may be on the beach. See them follow fishing boats. Gulls are good scavengers (SKAV-en-jers).

When your casual interest develops into a passion to see more of these wonderful birds, you will find that a pair of binoculars is a necessity. But even with your eyes alone, there is a great deal to see and learn about birds. Your eyes can pick out many birds and their actions along the way.

Birds commonly found in cities

Bird	Guide for identification	Song sounds like	Remarks
Robin	Gray back, rusty breast. Young have speckled breasts	"cheer-up" or "cheerily" (repeated)	When it stops moving on the ground it quickly stands up attentively.
English sparrow	Male has black throat, gray cap, white cheeks. Female does not have black throat.	Unmusical chirps	They always hop, never walk.
Starling	Purple and green sheen. Sprinkled with white dots. Looks black from distance. Short tail. Yellow bill in spring. Walks with a waddle.	Whistles, chatters	Congregate in noisy groups.
Catbird	Neat, dark gray, black cap.	Catlike mew	Also has many musical phrases.
Crow	Large, glossy, black.	"Caw, caw, caw"	Has deep, steady wing beat.

Bird	Guide for identification	Song sounds like	Remarks
Blue jay	Blue and white with crest. Fan-shaped tail.	Loud "jay, jay, jay"	It has many calls; some imitate other birds. Sassy, energetic. Eats eggs of smaller birds.
Slate-colored junco	Slate gray, white belly, white outer tail feathers, pink bill.	Light twittering sounds	Usually feeds on the ground.
Purple grackle	Long tail. Glossy purple or bronze.	Hoarse grating sound	Walks with graceful steps. Found in groups.
Chipping sparrow	Reddish cap, white stripe over eye, black bill.	Musical trill in one pitch only	Feeds mostly on the ground.
Towhee	Size of robin. Black top, rusty sides, white breast, white tail spots. Female same, but brown top.	"Chewink" "Drink your teeeeee"	Likes to feed in dead leaves and brushy places.

Bird	Guide for identification	Song sounds like	Remarks
Black-capped chickadee	Small bird, black cap, black throat, white cheeks.	"Chick-a-dee-dee-dee." Also "fee-bee-ee"	Friendly little acrobats. Will swing head down from branch. Responds to your imitated whistles.
Song sparrow	Heavily streaked dark brown breast with center spot.	Three deliberate notes, then hurried trills	One of the first birds to sing in the spring.
Cardinal	Red bird with crest, red bill. Female brownish.	Clear whistles: "whoit, whoit, whoit"	Loves sunflower seeds. Does not migrate.
Swallows	These birds fly back and forth, over fields and ponds, frequently changing direction and speed as they snatch flying insects.	"Twit-twit" "Che-veet"	Barn swallows have forked tails. Tree swallows have pure white under parts.

Bird	Guide for identification	Song sounds like	Remarks
Mallard duck	Male, grayish with green head and neck. White neck ring. Female, mottled brown.	Loud, short quacks	The most frequently seen duck in parks and ponds.
Downy woodpecker	White strip down back, under parts white, rest spotted and checkered black and white.	Little squealing notes: "peak, peak"	A small edition of a hairy woodpecker.
Flicker	A woodpecker with a brown back and white rump visible as bird flies. Heavy round spots on breast. Black crescent on breast. Red patch back of head.	Series of loud "wick wick, wick," etc.	Flicker often feeds on ground.

STARLING SILHOUETTE OVERHEAD

More to find out

1. On your walks you will find spots where certain birds are always found. Try to discover the reason for this: shelter, water, special food, etc.
2. Try to find nesting birds. Do not disturb them.
3. Keep records when you see birds starting to nest. Include the number of eggs, length of incubation, number of young, feeding records, date young left nest, etc.
4. Which birds in your neighborhood are winter residents?
5. Carefully examine a parakeet. Study its warmth, rapid heart beat, eyelids, ears, nostrils, tongue, toes, down feathers, eating habits, method of preening.
6. Build a feeding platform, bird house, bird bath, or suet hanger in your back yard. You can study wild birds more closely this way.
7. Join a bird club. You will get plenty of fresh air, fun, exercise, and companionship, as well as a fascinating hobby!
8. Why do birds migrate?
9. Visit someone who raises pigeons.
10. Observe how a bird slopes its body to one side when making turns. This is called banking. Compare with airplanes.

GARBAGE-CAN-COVER BIRDBATH

THE SKY IN THE DAYTIME

When you leave your home for a walk on a clear day, you are greeted by the beautiful blue sky. There are many interesting things to know and observe about the sky.

Why is the sky blue? It sounds like a simple question that children often ask. Yet, if you look in the *Encyclopaedia Britannica* under "sky" you will find that many top-notch scientists disagree upon the correct answer. Here is probably the most accepted explanation:

You know that sunlight is made up of all the colors of the rainbow. (Glass prisms, corners of fish tanks, or the beveled edges of mirrors often break up sunlight into bands of red, orange, yellow, green, and blue colors.) When sunlight shines on the earth, it passes through the atmosphere. The gases and particles of dust in the air scatter the light rays differently. Blue rays are scattered the most, red the least.

WINTER SHADOW

The reddish appearance of the sun at sunrise and sunset can also be explained. The red light of the sun can penetrate the atmosphere better than blue light. When the sun is low we see it through more air and dust than when it is overhead. This prevents the blue rays from reaching us. If there are clouds in the west we often get spectacular sunsets as the pink light is reflected from these clouds.

Do you know that you see the sun for about three minutes after it has actually set in the west? The atmosphere around the earth bends the light from the sun so that you still "see" the sun when it is below the horizon. This effect is called a *mirage* (mi-RAHZH).

It is interesting on your walks to see where the sun rises and sets at different times of the year by using buildings or other landmarks as reference points. On or about December 21, the first day of winter, the sun rises south of east. It makes the lowest arc of the year through the sky and sets south of west. The first day of winter is also the shortest day of the year.

SUMMER SHADOW

About June 21, the first day of summer, the sun rises north of east and makes a high arc through the sky. It sets north of west. This is the longest day of the year.

On or about September 23 and March 21, the beginning of fall and spring, the sun rises exactly in the east and sets exactly in the west.

Keep your eyes open for this yearly path of the sun in the sky. It is caused by the fact that the earth's axis is tilted 23½ degrees all the time that it is revolving around the sun.

Clouds can be seen in the sky most of the time. Besides their interesting formations, they may also help you predict the weather. But first here are certain facts you may wish to know.

When water evaporates from oceans, lakes, rivers, land, and thousands of other places, it turns into invisible water vapor. Upward air currents sometimes carry it high into the air. When the water vapor is cooled it condenses into water on tiny bits of dust. The extremely small droplets are not heavy enough to fall to the earth. This suspended mass of tiny droplets is called a cloud.

Clouds often change their shapes while you look at them. One reason is that it is windy up in the air and parts of the clouds are easily blown away. You can see them scudding about. You may notice also that the winds frequently blow clouds in opposite directions. Often the wind direction on the ground may be completely different from that up in the air.

Have you ever seen an entire cloud, or part of one, appear or disappear very quickly? A sudden warm, dry wind will cause a cloud to evaporate. On the other hand, warm moist air may be suddenly cooled and form a cloud. Very high clouds may have tiny ice crystals in them instead of water droplets. Parts of these clouds may appear and disappear very rapidly as the ice crystals melt and refreeze, meeting warm or cold winds.

The color of a cloud depends upon the position of the sun. When the sun is low in the sky a thin cloud overhead may have its base outlined sharply. The same cloud may hardly be visible when the sun is shining directly through it. The thickness of a cloud and the size of the droplets may also cause different shades of gray.

The drops in a cloud combine to form bigger drops. They remain suspended in the air mainly because of rising air currents. When they become heavy enough they will fall as rain. Rain clouds are quite dark. However, all dark clouds are not necessarily rain clouds.

Sometimes on a foggy day, you may be walking in a cloud. That is because a fog is a cloud resting on the ground.

There are three main types of clouds:

1. *Cirrus* (SIH-rus) *clouds.* These thin, wispy clouds, resembling feathers, are composed of ice crystals. They are about six miles above the earth. They are commonly known as "mares' tails" or "witches' brooms." They usually mean rain or snow in two or three days, depending upon the season.

2. *Cumulus* (CYOOM-yoo-lus) *clouds.* These scattered clouds look like fluffy balls of cotton. They have flat bases and rounded upper surfaces. They cast shadows on the earth. They are considered fair-weather clouds.

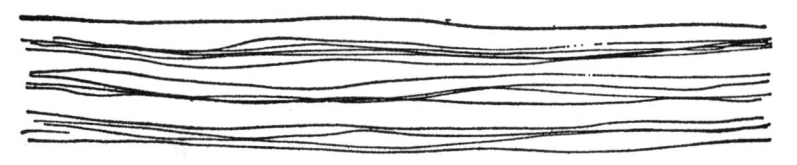

3. *Stratus* (STRAY-tus) *clouds.* These cover the entire sky like one continuous gray sheet, hiding the sun. Clouds like these may produce a drizzle.

There are many combinations of these three types of clouds, depending upon the altitude at which they are formed.

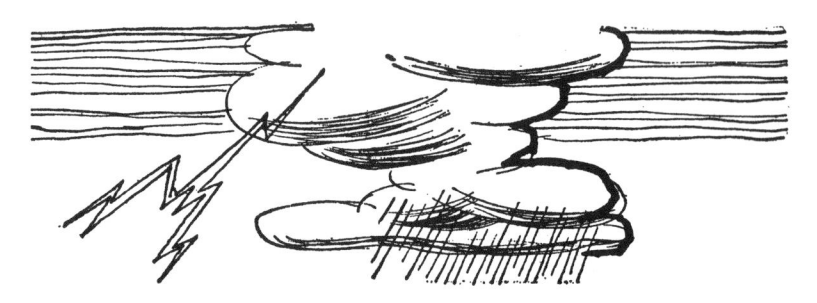

A *thunderhead* is a cumulus cloud which grows rapidly into a tall anvil-shaped, dark cloud. It usually occurs on a warm summer day. When you see this warning, start heading for home or another shelter because a lightning storm may be upon you very quickly.

Lightning is a gigantic flash caused by a discharge of static electricity. The electricity jumps between two clouds, or from a cloud to the earth and from the earth to a cloud. It is similar to the tiny flash you make when you scuff your feet on a woolen rug and then hold your finger near a radiator or a door knob.

Thunder is caused by the heating of the air by lightning. When air is heated, of course, it expands. And as the air expands violently it produces the familiar sounds of thunder. We see the lightning almost immediately, but the sound travels to us at the rate of one-fifth of a mile per second. That is why thunder always comes to us after the lightning.

If you are caught out of doors during a lightning storm, stay away from any tall tree which is apart from others. Lightning usually strikes the highest object near it. Also stay away from metal fences, because metals conduct electricity.

Have you ever heard a weather report which referred to the ceiling and also the visibility in a certain area?

The *ceiling* is the height from the ground to the bottom of the lowest cloud. Weathermen find this distance by releasing large helium balloons and following them with surveyors' telescopes. At night, strong vertical searchlights shine a beam to the bottom of a cloud. Instruments measure this height.

Ceiling unlimited means that the lowest clouds are over 10,000 feet high and that less than half the sky is cloudy. When the ceiling is less than 50 feet, because of fog or dust, it is reported as *ceiling zero*.

Visability means the greatest distance that objects can be seen by a person on the ground. When the visibility is less than a mile, no airplanes are allowed to take off or land at an airport.

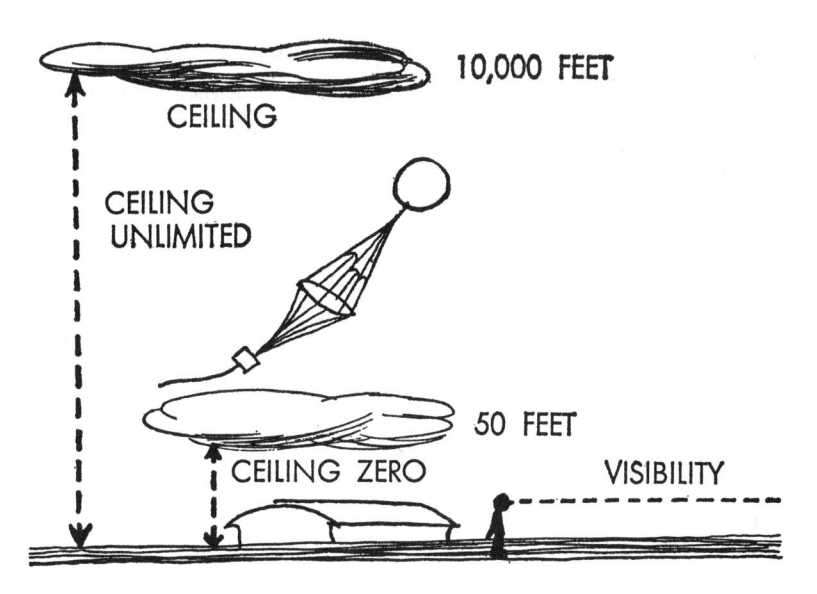

Compare your estimates of ceiling and visibility with the official ones.

When water vapor condenses above 32° Fahrenheit, *rain* is formed. Sometimes the rain falls through very cold air on the way down and freezes into little balls of ice. This is called *sleet*.

Hail is formed in another way. Sometimes during a thunderstorm strong upward winds carry the raindrops high into the air, where the drops freeze. When they fall through the clouds more water condenses around these bits of ice. Now the upward wind carries them up again and a new layer of ice is formed around the old. Again and again this process is repeated, until the weight of the hailstones becomes more than the upward wind can lift. Then they fall to the ground. Sometimes they can be as large as baseballs. Most hailstones, however, are small as peas.

Hail occurs mainly in the summertime. The balls of ice are of widely different sizes. Sleet looks like round beads of ice which are all about the same size. Sleet usually occurs in cold weather.

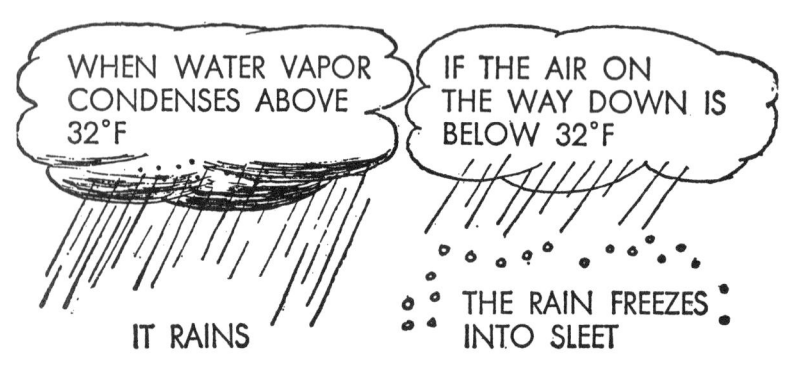

WHEN WATER VAPOR CONDENSES ABOVE 32°F
IT RAINS

IF THE AIR ON THE WAY DOWN IS BELOW 32°F
THE RAIN FREEZES INTO SLEET

Snow is formed when the water vapor in the clouds is cooled below 32° Fahrenheit. Snowflakes are six-sided crystals. There are thousands of different patterns, but no two snowflakes are exactly alike. Use a hand lens to quickly examine one which falls on your sleeve.

During a walk after a shower, you may see a rainbow up in the sky. It is caused by the breaking up of light by tiny drops of water, still suspended in the air after the recent rain. Observe that the colors form bands of red, orange, yellow, green, blue, and violet. The red is on the top and the violet on the bottom. Try to see a fainter rainbow on top of the first one with the colors reversed—the red is on the bottom this time.

In order for you to see a rainbow, the sun must be behind you. That is why rainbows are usually seen early or late in the day.

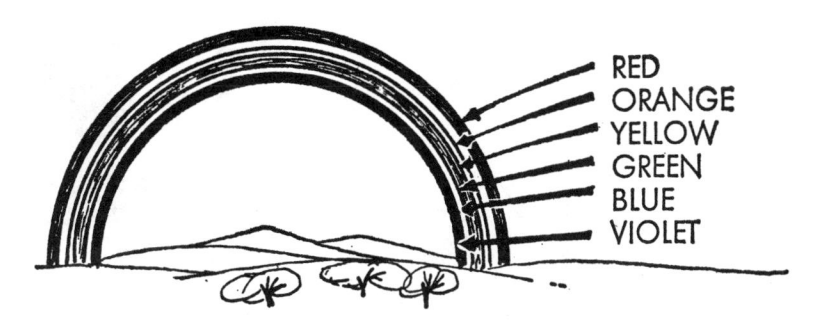

RED
ORANGE
YELLOW
GREEN
BLUE
VIOLET

More to find out

1. Make a rainbow by using the fine spray of a garden hose. See how the sun must be behind you.
2. Sometimes airplanes leave "vapor trails" in the atmosphere. What causes this?
3. Make a fog in a glass milk bottle. Fill a bottle with warm water. Pour out all but a half inch of water. Place an ice cube on top of the open bottle. Explain why a cloud forms.
4. How can you tell how many miles away a lightning flash occurred?
5. What is meant by "cloud seeding?"

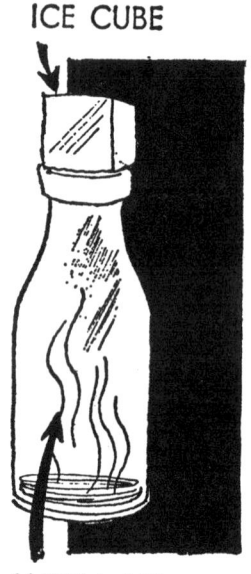

ICE CUBE

WARM AIR

OBSERVING THE NIGHT SKY

A walk with a companion at night has a particular charm. Gazing upward at the starry sky gives one a sensation of awe—almost a religious feeling. Sizes and distances are so vast and overpowering. How puny and insignificant people seem to be!

Let us start our observations with our nearest neighbor, the moon.

When the moon rises in the east as a full moon it seems to be about the same size as the sun, but don't let this optical illusion fool you. The moon is only about 2,000 miles across its face. The diameter of the sun is over 400 times as large. The reason for the moon's large appearance is that it is hundreds of times closer than the sun.

The moon is within practical rocket distance—about

143

240,000 miles away. By comparison, the sun is about 93,000,000 miles away.

The moon sets in the west, just as the sun does. This is caused by the spinning of the earth from west to east. Because the moon also revolves around the earth, it sets about 51 minutes later every day.

The moon does not give off any light of its own. We see it only because it reflects back to us the sunlight which is shining upon it.

What is the Man in the Moon? The eyes, nose, and mouth are flat places on the moon's surface. They look darker than other parts. Many years ago people thought these were bodies of water and called them "seas." Actually, the so-called seas are flat plains.

If you look at the moon even with medium-power binoculars, you will see many mountains and craters as well as the plains. You will also see rays which seem to spread out from craters. These have never been satisfactorily explained by astronomers.

The moon's surface can be seen so clearly because there is no atmosphere on the moon.

During a month we see different shapes or phases of the moon. Sometimes it is a crescent, at other times it is a round silvery ball. Sometimes we cannot see the moon at all.

The reason for the moon's changing appearances is that only one half of the moon is illuminated by the sun, and as the moon revolves around the earth once a month it is in different positions with respect to us and the sun. When we see the entire illuminated side, we call it a full moon. But when we cannot see the illuminated part at all we call it a new moon. In between these two positions there are the other shapes.

THE HORNS OF THE CRESCENT MOON POINT AWAY FROM THE SUN

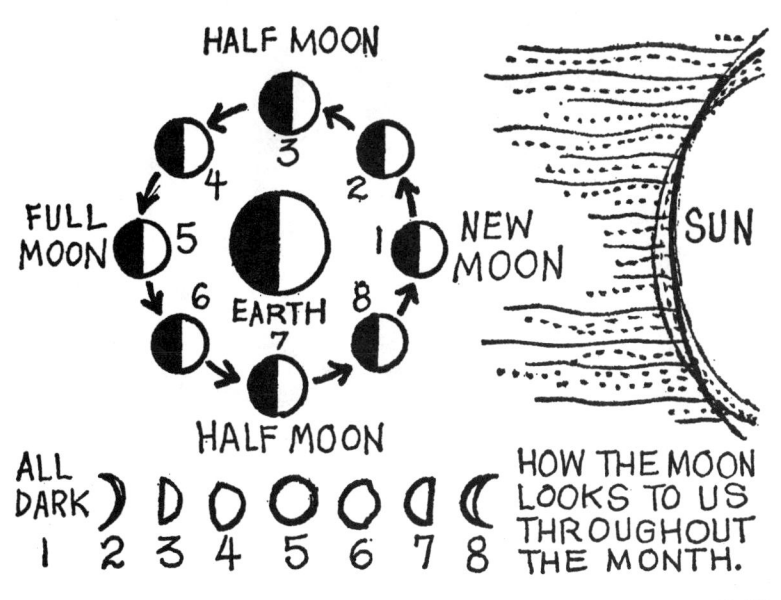

HOW THE MOON LOOKS TO US THROUGHOUT THE MONTH.

145

When the moon is new and only a thin crescent is seen, look for it low in the western sky. In one or two hours after the sun sets, it vanishes over the horizon.

If you look carefully at the crescent moon you may see the circular outline of the dark section of the moon. This is always interesting to observe. It is caused by the lighted half of the earth reflecting light back to the dark side of the moon. It is called *earth shine*.

The horns of a crescent moon always point away from the sun. You can tell exactly where the sun is by connecting a line to the two ends of the horns. Using the bow thus formed, shoot an arrow into the distant sun.

The growing or waxing half-moon is also called the first quarter. It rises in the east about noon. It reaches its highest point at sunset and sets near midnight.

The full moon rises in the east just as the sun sets in the west. It is in the sky all night.

It looks orange when it is low on the horizon because we see it through more dust and atmosphere. Red light rays penetrate the atmosphere better.

When the full moon first comes up it looks much bigger than when it is overhead. By a simple test, however, you can prove that it is the same size.

Hold a pencil with an eraser on it in a vertical position at arm's length. See how the eraser covers the moon when it is rising. Do the same when the moon is overhead and compare the two observations.

The third quarter or waning half-moon rises about midnight. It is highest at dawn and sets about noon. You can see it in the morning in the west. Its round side is toward the east. Many people are surprised to see the moon in the daytime.

By the way, can you see what is wrong with this picture?

You are right if you say that the star cannot be seen because the moon is in the way.

If you ever see a halo around the moon, it is caused by the reflection of moonlight from ice crystals high up in the atmosphere.

Stargazing is one of the most fascinating hobbies. However, it takes time to develop the ability to look at the sky and recognize stars and constellations like old friends. By learning a few at a time you can prepare yourself for a lifetime of pleasure.

Astronomers tell us that there are many millions of stars. But without the aid of a telescope you can only see between two and three thousand at one time.

Stars are huge balls of hot gases like our sun. You have learned that our sun is 93 million miles away, but the nearest star is about 25 trillion miles away. Most stars are many thousands of times as far.

When you look at a star, try to remember that its light may have been traveling through space for many years. You are seeing it as it used to be. Light travels at 186,-000 miles per second. The distance light travels in one year is called a *light year*. After our sun, the nearest star, Alpha Centaure (AL-fa sent-AW-ry), is 4⅓ light years away. This is a dim star and can only be seen in the southern hemisphere. If you live in the northern hemisphere, the nearest bright star is Sirius (SEER-ee-us), the Dog Star. It is 8.6 light years away.

Some stars look brighter than others. This does not mean that they are larger. Small stars may appear larger and brighter than others because they are closer.

If you look carefully you will notice that the stars are not alike in color. Some look bluish, others are yellow.

You may also see some that are green or red. Astronomers use instruments to measure the temperature of stars. They have discovered that red stars are much cooler than white or bluish ones. A yellow star, like our sun, has a medium temperature compared with other stars.

Stars seem to twinkle because you are looking at them through atmosphere which is constantly moving.

There are stars in all directions from the earth. During the day, the strong sunlight brightens the sky and prevents us from seeing the much dimmer stars. However, balloonists and high-flying jet pilots see the stars in the daytime. As they gain altitude there is less atmosphere to reflect the sunlight and the sky appears black.

ABOVE THE EARTH'S ATMOSPHERE THE SKY IS BLACK AND THE STARS ARE ALWAYS VISIBLE

The stars seem to remain in the same place in relation to each other. Actually, all stars are moving in different directions at fantastic speeds. They have been doing this for millions of years. They appear to be fixed in one position because they are so far away that their changes are not noticed for many hundreds of years.

If you use your imagination certain stars will seem to form the outlines of pictures. A pattern made by a group of stars is called a *constellation.* Ancient people gave these constellations interesting Latin names and made up stories about them, which have persisted to this very day.

Very few constellations look like their names. With a little practice, however, you can learn to pick them out in the sky. You can also name the most prominent stars in the constellations.

Some constellations are shown on these pages. The library has many books which are devoted entirely to stars. Perhaps a friend who knows the stars will start you off on this hobby some night.

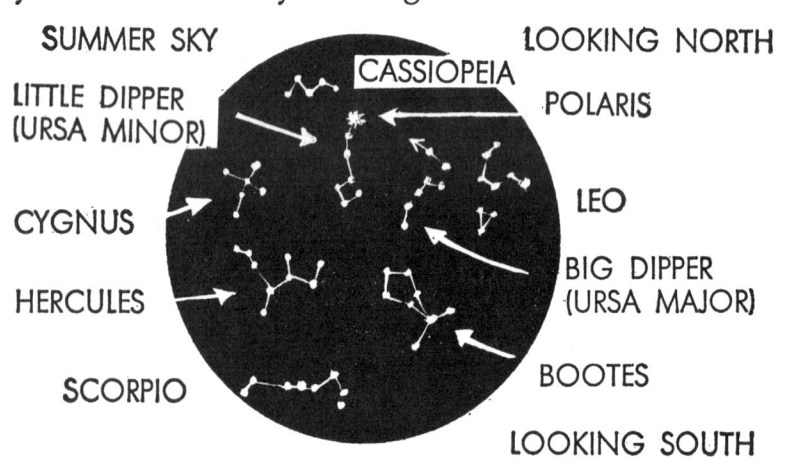

SUMMER SKY
LOOKING NORTH
LITTLE DIPPER (URSA MINOR)
CASSIOPEIA
POLARIS
CYGNUS
LEO
HERCULES
BIG DIPPER (URSA MAJOR)
SCORPIO
BOOTES
LOOKING SOUTH

Most of the stars seem to move toward the west every night. They do this without changing their positions in relation to each other. This gigantic optical illusion occurs because the earth itself is rotating from west to east.

There is one star that does not seem to move in the sky. It is the North Star, also called the Pole Star or Polaris (poh-LAR-is). The axis of the turning earth points to it in space. It is visible in the northern hemisphere all through the night.

To understand this better, try to imagine yourself riding on a carousel at a carnival. As you look away from the merry-go-round the spectators, booths, and scenery move past you in an opposite direction to your motion. Some parts of the scenery are completely lost from view while you are turning to the other side. Now look up and you will see the ceiling, or other things above you all the time you go around. The place where the center shaft of the merry-go-round pierces the ceiling does not seem to make a circle at all.

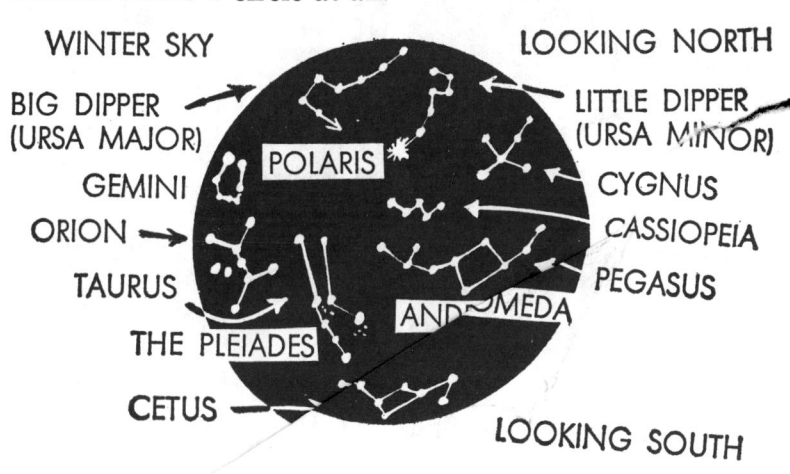

WINTER SKY — LOOKING NORTH

BIG DIPPER (URSA MAJOR)
LITTLE DIPPER (URSA MINOR)
POLARIS
GEMINI
CYGNUS
ORION
CASSIOPEIA
TAURUS
PEGASUS
ANDROMEDA
THE PLEIADES
CETUS

LOOKING SOUTH

The North Star happens to be almost in the spot where the axis of the earth would pierce the sky. Because it remains in one spot and is not affected by the rotation of the earth, it is used by navigators for telling direction.

The stars close to it do not set on the western horizon. On clear nights they can be seen all the time. They are called *circumpolar* stars. They seem to make a complete circle of 360 degrees in 24 hours. Therefore they move $\frac{360}{24} = 15$ degrees in one hour. Perhaps you can devise a way of telling time by watching the stars.

The North Star is not the brightest star in the sky, as many inexperienced people think. It can be seen clearly once you know how to find it and recognize it.

To find it for the first time, use a compass and face north. Point a finger halfway between the horizon and overhead. You will be pointing toward the North Star.

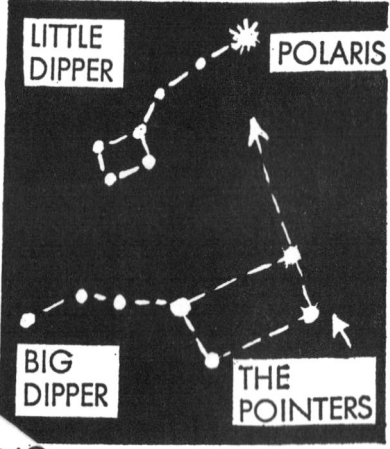

TWO WAYS OF FINDING ⁻ NORTH STAR

The other method, which is widely used, is to look for the Big Dipper in the northern sky. This is easily found. Make a line between the two end stars of the bowl of the Dipper. These two stars are called the Pointers. Continue this line for about five times the distance between the Pointers. It will bring you to the North Star. The North Star is also at the end of the Little Dipper.

The stars are all around our earth in every direction. Of course, we can only see them when it is dark. During an eclipse of the sun the stars are visible in the daytime. Because we revolve around the sun once a year, the night sky in the summer is not the same as the night sky in the winter. That is why the summer constellations that we see are not the same as the winter constellations. Of course, those very close to the North Star can be seen all year.

Some bright objects in the sky look like stars but they are planets. Like the moon, they reflect light from the sun. We can often see Venus, Jupiter, Mars, and Saturn with the naked eye—if we know just where to look for them. Sometimes a planet is so bright that it becomes visible before the stars come out. It is then called an evening star.

Planets are called "wanderers." Because each moves in its orbit around the sun, they appear to move among the constellations. Of course they are much nearer than the constellations. The movement of planets with respect to the constellations is hardly noticeable from one night to another.

Venus is the brightest object in the sky, after the sun and the moon. If you know it is in the sky you will easily find it.

Mars is also easy to recognize because of its reddish color. Jupiter looks bright like Venus and is often mistaken for it. However, Venus always sets soon after the sun. Jupiter stays out later than Venus.

Do not try to see the rings on Saturn, even with binoculars. You must use a telescope to see them. Even a small one will do.

You can find out which planets can be seen at certain times by referring to *The Science News Letter* (the last issue of the month). *Natural History,* a magazine published by the Museum of Natural History in New York City, also contains this information. Certain newspapers and magazines carry monthly star maps and planet guides. Almanacs also have charts. Your librarian certainly knows how to put her finger on this information.

During one of your evening walks you may see a fiery streak across the sky. It may be a "shooting star." It is not really a star, but a chunk of material called a *meteor* from somewhere out in space. It is attracted by the earth's gravity. As it falls at a very high speed through our atmosphere it gets heated to about 4000° Fahrenheit by the friction against the air. Sometimes it does not get completely burned up and a piece may fall to the ground. It is then referred to as a *meteorite.*

Some meteors travel in groups and are more numerous on certain nights of the year.

More to find out

1. What is the Milky Way? What is a galaxy?
2. What are the Northern Lights that sometimes flash and color the night sky?
3. Take pictures of "star trails." On a dark night, when there is no moon, prop up a loaded camera so it points to the North Star. Open the shutter. Set distance for infinity. Use a fast film. Keep shutter open for at least two hours.
4. Try to look for one of the artificial satellites as it passes through the sky. Get the time from the newspaper or a planetarium.
5. Find out more about clusters of meteors called Perseids, Leonids, and Geminids.
6. Take time-exposure pictures of the moon.
7. Why is it hard to see stars over a city?
8. Visit a planetarium.

Some books which will help you become a more observant walker

George Barr, *Science Research Experiments for Young People*, New York: McGraw-Hill Book Company, Inc., 1958; Dover Publications, Inc., 1989

George Barr, *Young Scientist Takes a Ride*, New York: McGraw-Hill Book Company, Inc., 1960

George Barr, *More Research Ideas for Young Scientists*, New York: McGraw-Hill Book Company, Inc., 1961

George Barr, *Sports Science for Young People*, New York: McGraw-Hill Book Company, Inc., 1962; Dover Publications, Inc., 1990

George Barr, *Young Scientist Looks at Skyscrapers*, New York: McGraw-Hill Book Company, Inc., 1963

George Barr, *Science Projects for Young People*, New York: McGraw-Hill Book Company, Inc., 1964; Dover Publications, Inc., 1986

George Barr, *Showtime for Young Scientists*, New York: McGraw-Hill Book Company, Inc., 1965

George Barr, *Young Scientist and the Fire Department*, New York: McGraw-Hill Book Company, Inc., 1966

William H. Crouse, *Understanding Science*, New York: McGraw-Hill Book Company, Inc., 1956

Alan Devoe, *This Fascinating Animal World*, New York: McGraw-Hill Book Company, Inc., 1951

Bessie M. Hecht, *All about Snakes*, New York: Random House, Inc., 1956

William Hillcourt, *Field Book of Nature Activities*, New York: G. P. Putnam's Sons, 1950

E. L. Jordan, *Hammond's Guide to Nature Hobbies*, Maplewood, New Jersey: C. G. Hammond and Company, 1953

Richard R. Kinney, *Guide to Gardening with Young People*, Englewood Cliffs, New Jersey: Prentice-Hall, Inc., 1955

Robert S. Lemmon, *All about Birds*, New York: Random House, Inc., 1955

Robert S. Lemmon, *All about Moths and Butterflies*, New York: Random House, Inc., 1956

Kelvin McKready, *A Beginner's Guide to the Stars*, New York: G. P. Putnam's Sons, 1945

Roger Tory Peterson, *A Field Guide to the Birds*, New York: Houghton Mifflin Company, 1947

Richard H. Pough, *Audubon Bird Guides*, Garden City, New York: Doubleday & Company, Inc., 1953

H. A. Rey, *Find the Constellations*, Boston: Houghton Mifflin Company, 1954

Herman Schneider, *Everyday Machines and How They Work*, New York: McGraw-Hill Book Company, Inc., 1950

Herman Schneider, *Everyday Weather and How It Works*, New York: McGraw-Hill Book Company, Inc., 1951

Herman and Nina Schneider, *Rocks, Rivers and the Changing Earth*, New York: William R. Scott, Inc., 1952

Julius Schwartz, *It's Fun to Know Why*, New York: McGraw-Hill Book Company, Inc., 1952

Julius Schwartz, *Through the Magnifying Glass*, New York: McGraw-Hill Book Company, Inc., 1954

Edwin Way Teale, *The Junior Book of Insects*, New York: E. P. Dutton & Company, Inc., 1953

Rose Wyler and Gerald Ames, *The Golden Book of Astronomy*, New York: Simon and Schuster, Inc., 1955

Herbert S. Zim, *Frogs and Toads*, New York: William Morrow & Company, Inc., 1954

Herbert S. Zim, *Trees*, New York: Simon & Schuster, Inc., 1953

Herbert S. Zim and Clarence Cottam, *Insects: A Guide to Familiar American Insects*, New York: Simon & Schuster, Inc.

Herbert S. Zim and Donald F. Hoffmeister, *Mammals*, New York: Simon & Schuster, Inc., 1955

INDEX

Airplane, 55
Alloys, 51, 53, 57, 60, 61
Aluminum, 55, 56
Antifreeze, 72
Ants, 112–113
Asbestos, 48
Asphalt, 11
Asphaltum paint, 55

Banked roads, 15
Basalt, 47
Bat, 106
Bees, 115–116
Big Dipper, 153
Bird records, 132
Birds, 119–132
Blue jay, 129
Bluestone, 47
Brass, 57, 60
Bricks, 49
Bronze, 57
Brownstone house, 46
Butterflies, 108

Calcium carbonate, 44
Carbon dioxide, 33
Carbon monoxide, 33, 36
Cardinal, 130
Cast iron, 54
Catbird, 128
Caterpillar, 109
Cats, 100–101
Caves, 50
Ceiling, 139
Cement, 12
Cement block, 49
Cemetery headstone, 41, 44, 63
Centipede, 117
Chickadee, 130
Chickweed, 90
Chipmunks, 102–103
Chipping sparrow, 129

Chlorine, 37
Chlorophyll, 80, 81, 92, 93
Chromium plating, 56, 60
Chrysalis, 109
Cicada, 116
Cinder block, 49
Circumpolar stars, 152
Clay, 45, 68
Clouds, 135–138, 142
Coal, 36, 48
Cobblestones, 41, 47
Cocoon, 109
Compass, 53
Concrete, 12, 13, 72
Constellation, 150
Copper, 56, 57
Creosote, 17, 71
Crow, 124, 128
Crown of road, 10
Curbstone, 10, 14, 41, 47
Cut-off valves for house, 21

Dandelions, 89
Decoy, 71
Dogs, 37, 98–99
Doppler effect, 26
Downy woodpecker, 131
Duck, 131
Dust on dirt roads, 21, 37

Electric pole, 16, 17
English sparrow, 125, 128
Erosion, 62–72
Excavation, 66

Feldspar, 42
Fence, 58, 69, 71
Fern, 91, 95
Fire alarm boxes, 19
Firefly, 114
Flagstones, 45
Flicker, 131

158

Flies, 111
Flowers, 35, 86, 87, 95
Fog, 136, 142
Fossils, 44
Foundations of buildings, 49
Frogs, 105
Fuel oil delivery pipe, 21
Fungus, 92, 93

Galvanized iron, 52, 54
Gardens, 85, 87
Gas, 36, 76
Gasoline, 33
Glass, 49
Gneiss, 47
Gold-leaf, 61
Gold paint, 57
Grackle, 129
Granite, 14, 40
Grass, 67, 68, 90, 93
Grasshopper, 112
Gravel, 13
Gull, 127
Gully, 65, 66
Gutter, 10, 65
Guy wire, 19

Hail, 140
Hawk, 127
Hay fever, 91, 96
Heat and cold, 63, 64
Hedge, 85
High voltage, 17, 18
Hornblende, 42
Horses, 104

Ice, 64, 72
Igneous rock, 41, 47
Inertia, 14
Insects, 37, 76, 93, 107–118
Insulating material, 56
Insulators, 17
 on poles, 49
Iron, 52–55

Junco, 129

Lawn, 67, 89, 90
Lead, 58, 59
Leaves, 75, 80, 81

Lichens, 93
Light year, 148
Lightning, 83, 138, 142
Limestone, 12, 43, 44, 53

Macadam, 12
Mammals, 98
Manhole covers, 20, 21
Marble, 42, 43
Mercury, 61
Metals, 51–61
Metamorphic rock, 43, 45, 47
Metamorphosis, 108
Mica, 42, 50
Mica schist, 47
Millipede, 117
Mirage, 15, 134
Mold, 92, 96
Monuments, 41
Moon, 143–148
Mosquito, 110
Moths, 108
Mufflers, 24
Mushroom, 92

Nectar, 35
North star, 151–152

Odors, 31–39
Oil stains on roads, 21

Painted signs, 70
Paving of street, 11
Petals, 35
Pheasant, 126
Photosynthesis, 80
Pigeon, 125
Planets, 153, 154
Plantain, 90
Plants, 84–96
Poison ivy, 88, 96
Pollen, 35, 91
Power house, 18
Praying mantis, 114
Pruning, 75
Puddles, 68

Quartz, 42, 48

159

Rabbit, 103
Ragweed, 90
Railroad tracks, 28
Rain, 64–68, 136, 140
Rainbow, 141, 142
Resonance, 29
Road, 63, 64
Robin, 125, 128
Rocks, 64, 72
Roofs, 45, 48, 49, 57, 72
Roots, 67, 70, 80, 83
Rot, 17
Rust, 48, 52, 55, 57, 62, 69, 70

Salamanders, 105
Sand, 21, 48
Sandstone, 46, 47
Sedimentary rock, 45
Seeds, 96
Sewer pipes, 10, 21, 49, 55, 58
Shale, 12, 45
Shooting star, 154
Sidewalks, 46, 63, 67
Sirius, 148
Skidding, 21
Skunks, 104
Sky, blue, 133
Slate, 45
Sleet, 18, 140
Smog, 33
Smoke, 33
Snow, 141
Soil, 63, 81, 92
Solder, 58, 61
Song sparrow, 130
Sounds, 22–30
Spices, 39
Spider, 117
Squirrels, 101–102
Stalactites, 50
Stalagmites, 50
Star trails, 155
Starling, 125, 128
Stars, 148, 153

Statues, 57, 63
Steel, 52–55
Steeples, 57
Stone, 40, 68
Stones, artificial, 50
Storage batteries, 59
Store fronts, 49, 55
Street excavation, 36
Street lights, 19, 21
Street names, 20
Sunrise, 134–135
Swallow, 126, 130

Taste, 32
Taste and smell experiment, 39
Tern, 127
Termites, 17, 71, 113
Terrace, 67
Thunder, 138
Tin can, 59
Tin plate, 59
Tinfoil, 56
Tire screeching, 26
Toads, 105
Topsoil, 66, 67
Towhee, 129
Traffic signals, 19, 21, 57
Transformers, 18
Treasure hunt, 96
Trees, 35, 67, 70, 73–83

Vibration, 23
Vines, 88, 96
Visibility, 139

Warping, 68, 72
Water, 64, 68
Water strider, 115
Weeds, 71, 89–91
Wind, 35, 36, 63, 67, 72
Wires, 9, 18, 58
Woodpecker, 126

Zinc, 52

160